BARRIER TO THE BAYS

NUMBER THIRTY-FIVE
Gulf Coast Books
Sponsored by Texas A&M University–Corpus Christi
   Larry McKinney, General Editor
   John W. Tunnell Jr., Founding Editor

BARRIER TO THE BAYS

The Islands of the Texas Coastal Bend and Their Pass

MARY JO O'REAR

TEXAS A&M UNIVERSITY PRESS
College Station

COPYRIGHT © 2022 BY MARY JO O'REAR
All rights reserved
First edition

This paper meets the requirements of ANSI/NISO Z39.48-1992
(Permanence of Paper).
Binding materials have been chosen for durability.
Manufactured in the United States of America

A list of titles in this series is available at the end of the book.

LIBRARY OF CONGRESS CATALOGING-IN-PUBLICATION DATA

Names: O'Rear, Mary Jo, 1943– author.
Title: Barrier to the bays: the islands of the Coastal Bend and their pass / Mary Jo O'Rear.
Other titles: Gulf Coast books; no. 35.
Description: First edition. | College Station: Texas A&M University Press, [2022] | Series: Gulf Coast books; number thirty-five | Includes bibliographical references and index.
Identifiers: LCCN 2021041131 (print) | LCCN 2021041132 (ebook) | ISBN 9781623499402 (cloth) | ISBN 9781623499419 (ebook)
Subjects: LCSH: Barrier islands—Texas—Coastal Bend—History. | Bays—Texas—Coastal Bend—History. | Coastwise navigation—Texas—Coastal Bend—History. | Aransas Pass (Tex.: Strait)—History. | BISAC:HISTORY / United States / State & Local / Southwest (AZ, NM, OK, TX) | TECHNOLOGY & ENGINEERING / Civil / Flood Control
Classification: LCC F392.C517 O265 2022 (print) | LCC F392.C517 (ebook) | DDC 976.4/113—dc23
LC record available at https://lccn.loc.gov/2021041131
LC ebook record available at https://lccn.loc.gov/2021041132

To the stalwarts—

Robert Joseph Holoubek
Jessica Lee O'Rear
Marc Anthony Beban
Kenna Rose O'Rear Beban

CONTENTS

Acknowledgments ix

PART I. EARLY DAYS
    CHAPTER 1   Prehistory 3
    CHAPTER 2   Humans 7

PART II. EXPLOSIVE DAYS
    CHAPTER 3   Horse Marines 21
    CHAPTER 4   Sad Havoc 27
    CHAPTER 5   Harbingers 35
    CHAPTER 6   Hellfire 43

PART III. ENTREPRENEURIAL DAYS
    CHAPTER 7   Herders and Hiders 57
    CHAPTER 8   Helmsmen 63
    CHAPTER 9   Harvesters 76
    CHAPTER 10   Hustling with Hope 83
    CHAPTER 11   Hard Tracks and Haupt 88
    CHAPTER 12   Hydrodynamics and Dynamite 99

PART IV. ENTERPRISE DAYS
    CHAPTER 13   Harbor Home 111
    CHAPTER 14   Hunters and Hard Hulls 121
    CHAPTER 15   Hurricanes 132

PART V. EXPANSION DAYS

CHAPTER 16　Hall's Bayou Enhanced　151
CHAPTER 17　Holding Firm　161
CHAPTER 18　Hawsepipers　169
CHAPTER 19　Henry's Goodbye　175

Notes　181
Bibliography　223
Index　239

ACKNOWLEDGMENTS

Several years ago, when thanking the many who worked with me putting together *Bulwark against the Bay: The People of Corpus Christi and Their Seawall*, I singled out Robin Borglum Kennedy. Without her, that book would never have been completed. The same is true of two others with respect to *Barrier to the Bays: The Islands of the Texas Coastal Bend and Their Pass*. Tom Stewart created the most effective maps of this area available and has allowed me to adapt them to each chapter. Moreover, his writings on the barriers have been as basic to my research as the memorabilia he has shared. Mark Creighton gave me ideas, insight, and access to people, documents, and experiences unique to island life. He personifies the individuality and devotion of his fellow Port Aransans. Without either of them, this book would never have happened.

But they are not alone. Charles R. Cable, Greg Smith, and Ned and Sherrie Teller granted me precious time and interviews; Donald Chipman, Mike Campbell, and the late Wes Tunnell read early chapters; Sam Pullig and Robert Holoubek read later ones; and Allan Hayes joined Tom and Mark in proofing significant parts of this book. They shared their suggestions out of generosity of spirit; I alone am responsible for my mistakes and omissions.

Nothing could have been written without the previous work of Norman Delaney, Murphy Givens, and C. Herndon Williams; they stand as foremost historians of the Coastal Bend, as does J. Guthrie Ford, who delighted in chronicling the stories of his island. Jim Moloney contributed vital documents and graphics, as did Sam Pullig, Renato Ramirez, Murphy Givens, Carroll Scogin-Brincefield, Doug Kubicek, Rick Pratt, Richard Wright, Gary McKee, Ken Howell, George Cooper, Robin Borglum Kennedy, Monsignor Mike Howell, Roger Raney, Roy Smith, Dag

Nummedal, Linda Salitros, Cecilia Gutierrez Venable, Alan Lessoff, and out of the past, my late husband, James Lee O'Rear.

Particularly invaluable has been the help of Laura Garcia, Patricia Herrera, and Maria Carter in the Local History Room of the Corpus Christi Public Library, and of Cassidy Mickelson of the Texas Maritime Museum. The following contributed significantly as well: Carol Rehtmeyer of the Corpus Christi Museum of Science and History; Mark Creighton of the Port Aransas Museum (also known as PAPHA); Diana Cox and Susan Neal of Gilcrease Museum; Sara Pezzoni and Ben Huseman of the University of Texas at Arlington Libraries; Maria Adcock of the Victoria Regional History Center; and Allison Ehrlich of the *Corpus Christi Caller-Times*.

Inestimable patience marks Jay Dew as an editor of value; he and his Texas A&M University Press staff have made my every publishing adventure a joy.

The greatest thanks, however, go to my friends, who have tolerated my digressions and withdrawals and excitements with affectionate support, and to my family. San Antonio Pulligs and Louisiana Holoubeks, including those in Virginia, encompass some of the most enthusiastic history buffs in the nation; California O'Rears and Bebans comprise the most loving.

BARRIER TO THE BAYS

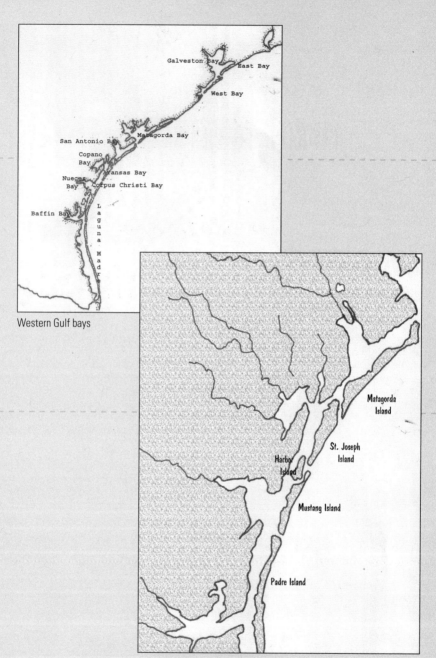

Western Gulf bays

Significant Gulf Coast islands

PART I. EARLY DAYS

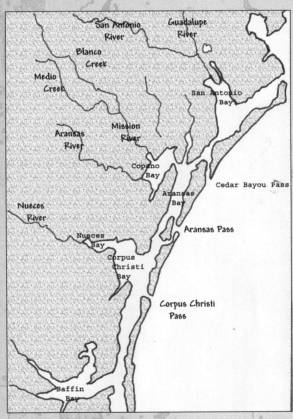

Natural Coastal Bend waterways

CHAPTER I

Prehistory

       The movement was slight at first, an imperceptible speck of dust propelling down the jagged edge of a Mesozoic cliff. Then more jots jumbled together, rolling into each other, clustering, tumbling, finally crashing as huge sand sheets into basins below, themselves old sedimentary seabeds. Thus were the Great Plains layered bit by bit, as schist and granites, ripped by wind from mountain rises, deposited themselves onto baseland bottoms. No less violently formed were the plains' cousins, the Gulf barrier islands, almost a million years later—with wind playing a far different role. But it was water that marked the real difference between the older plains formations and the emerging coastal isles. It started out frozen, vast glacial packs encasing mountaintops and solidifying straits, which began to melt around 8000 BCE. Freed by increasingly higher temperatures, seabound ice cracked apart into oceans and, in time, inundated continents. Land-bound ice thawed, warmed, then merged into gravity-propelled water chutes. Coursing from mountain rises, wresting gravel, sand, and silt in their wake and tossing them suspended in their flow, they raged, their power growing as creeks merged into streams and streams poured into rivers. But as they swept downward onto the shoreline of Texas, a vast flatland of mud slowed them. The more they slowed, the less suspended were their sediments, until the heaviest sank into encompassing marshes. But it was the final push through bays—bays themselves formed upon "drowned river valleys" of earlier eras—that ultimately emptied the rivers, thrusting them crossways into offshore currents. These swept remaining particulates into larger flows until, colliding into countercurrents or swept up by stronger tides, the particles dropped. Thus, sands that had coursed down from highland to seafloor began formation anew. Bit by bit, they piled upon themselves and began to solidify. Centuries passed, water levels changed, and the layers kept rising. By the time Hammurabi was codifying

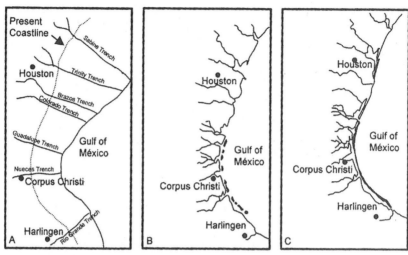

Diagrams showing the changes in the Texas coastline. The first indicates Texas about eighteen thousand years ago, the second reveals early island formations, and the third shows Texas at present. Reprinted from John W. Tunnell Jr. and Frank W. Judd, *The Laguna Madre of Texas and Tamaulipas*, with permission of Texas A&M University Press.

law in ancient Mesopotamia, Texas had lost its open Gulf—and gained a string of barrier isles.[1]

As well as an eternity of change. Silt accretion continued on each island, earliest layers now island core, and subsequent deposits rising vertically as shore-face formations. Atop them, waterborne sand and shells created beaches, and windborne particles dropped behind into hillocks and dunes. Some of these encased inlets, sodden with runoff and rising groundwater. Others, the larger dunes, sloped into grass-strewn mud banks that eased down into marshes or flats. From there, the isles sank into quiet lagoons or sounds extending parallel to the mainland. One, ranging south from Corpus Christi Bay to Rio Soto la Marina, was so placid that Spanish explorers christened it "Laguna Madre." Like its counterparts to the north—Aransas Bay, San Antonio Bay, Matagorda Bay, and East and West Bays off Galveston—this water body did more than separate islands from the shore. It cradled them with nutrients.[2]

Part of the nutrients came from northers that attacked during the winter. Their gusts destroyed weakened dunes and pushed larger ones westward, while wave surges forced seaborne silt onto beaches and beyond.

Prehistory

Another staple was hurricanes, cyclonic windstorms brewed in summer seas. Fueled by heated vapor rising aloft only to spiral downward in an ever-increasing onslaught of air masses, they swept across barrier isles, cut new troughs, and created more runnels in sediment-refreshed mudflats.[3]

But the most consistent sources were the longshore drifts that coursed around the islands during the year. During winter months, strong currents from upcoast carried particulates into the lagoons and the Gulf. Propelled by northwesterly winds, they flowed southward, dropping debris along the coastal shores of the barriers and on their beaches. During warmer months, offshore currents driven by southeasterly winds flowed northward until they too dropped residue. Over time, these deposits refurbished island bases, some layering weathered rock particles from the Edwards Plateau, others blending tourmaline from the plains, and a few spreading Rio Grande–wrenched basalt.[4]

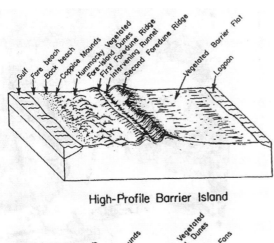

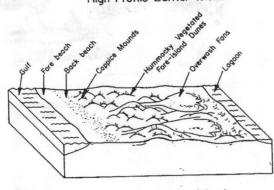

Profiles of the different types of barrier islands formed along the Texas coast. Courtesy of Dag Nummedal.

Dunes facing the Gulf. Courtesy of Jim Moloney.

But the offshore drifts provided more than base-building sediments; they also added shape. While heavier particles fell quickly along island cores during drift, lighter material coursed freely until drawn through gaps between the barriers. Funneled toward the open sea, they were compressed, slowed, and then dropped. Like their primordial ancestors, the particles fell downward, accumulated, and grew upward. This time, however, they emerged as island appendages, sandbars at passes and sand spits along barrier back sides. When humans approached around 1000 BCE, the barrier isles they saw featured crescent-shaped deltas, marsh-laden flats, and dunes latticed with vegetation.[5]

It was vegetation of profound variety; blue-green algae mats bordered lagoon shores, as did *cenicilla* and saltmarsh cordgrass. Inland runnels hosted cattails, and Gulf cordgrass grew on small sand mounds. Ocean-facing dunes sported morning glory vines, while sea oats clustered on their tops and seacoast bluestems graced their back sides. Also of intense diversity were the isles' animals—kangaroo rats, rice rats, gophers, and ground squirrels in the sand dunes; slider turtles, ribbon snakes, and rails around freshwater ponds and wetlands; skinks in vegetated flats; crabs and sea turtles on the beaches; gulls, terns, egrets, and great blue herons in the air. Below them and nurturing in the grasses and waters of the lagoons' estuaries were snails, clams, oysters, shrimp, whelks, black drum, and redfish.[6] During their growth off the Texas coast, the isles had become a cornucopia of life.

CHAPTER 2

Humans

But it was the estuaries and coastal wetlands that convinced hunters and fishermen to stay in the area around 800 BCE. Others had settled on the Coastal Bend during previous millennia, but the seas were too erratic to provide stability, and they had moved on. Not so the people of the Late Archaic Period.[1] The preponderance of black drum and spotted sea trout during the fall made shoreline fishing practical, and extensive oyster reefs provided shell weights for capture nets. Conches provided adzes and gouging tools, and lightning whelk and scallops became dietary staples. Even sea-oozed oil, washed onshore in gooey balls, turned everyday pottery into water-resistant jars. The proximity of the Gulf Prairies to the bays was an additional advantage; summers were spent hunting on them. Besides meat and skins, deer and buffalo provided bone shafts from which fishing spears could be fashioned.[2]

By the time these fishing spears evolved into bows and arrows, the people of the coastal estuaries had mastered the isles. Drawn by their apparent abundance but unable to easily wade the distance, the people created water vehicles, boats they hollowed out from oaks and then made fluid resistant.[3] Hesitant at first to set off from familiar shorelines, they poled themselves across the lagoons to the back sides of the isles, then sloshed ashore onto mudflats. There and inland among the hillocks, they found food: snails and clams easy to gather; frogs, toads, and baby alligators not too difficult to catch; rats and raccoons available to trap. Sea turtle nests on the beaches revealed eggs ready for plunder, and beds close by provided oysters for the taking. Shorebirds like egrets and cranes tested hunters' longbow accuracy, and schools of redfish challenged women's woven nets. Dependent on the seasonal bounty of the isles, the people studied them—the longshore drifts that pushed boats southward at certain times of the year and northward at others, the passes that swept dugouts seaward if paddlers were not careful,

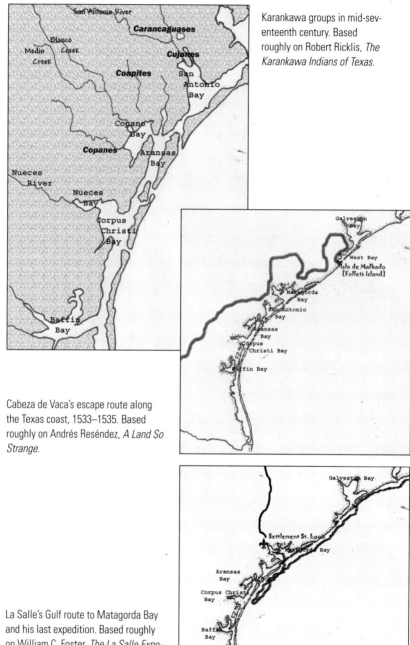

Karankawa groups in mid-seventeenth century. Based roughly on Robert Ricklis, *The Karankawa Indians of Texas*.

Cabeza de Vaca's escape route along the Texas coast, 1533–1535. Based roughly on Andrés Reséndez, *A Land So Strange*.

La Salle's Gulf route to Matagorda Bay and his last expedition. Based roughly on William C. Foster, *The La Salle Expedition to Texas*, and Robert Weddle, *The Wreck of the* Belle, *the Ruin of La Salle*.

Humans

Carancaguases, or Karankawas as they came to be known. Lino Sánchez y Tapia, *Indigenes du Mexico Voyage de J. Luis Berlandier*, plate 8, GM 4016.336, courtesy of Archives of Gilcrease Museum, Tulsa, OK.

the deltas and sandbars that changed shape every time a huge storm pushed through. Even the lagoons and channels merited study, as their increased salinity made gathering oysters harder as time wore on.[4]

For time favored these "wandering tribes" of the Central Texas coast. From their earliest arrival to the 1600s, they had marked it by island camping during the fall and inland roving during the summer. They had used it to measure territory, claiming Galveston's East Bay as their northernmost seasonal haven and Baffin Bay as their southernmost. They used time to plan movements onto the plains for prickly pear cacti and along the rivers for pecans. They developed language and, with time, divided themselves into groups—Carancaguases, Cocos, Cujanes, Coapites, and Copanes.[5] Most importantly, they took time to learn the waters around them. After years of observation, they knew the currents and shores of their islands well enough to give them names, although their designations were far different from the isles' final titles: Galveston, Matagorda, St. Joseph, Mustang, and Padre.[6] As consistent as the tides, the people of the coast made themselves part of the universe—until incompetents arrived.

They had not always been inept. The castaways struggling onto a tiny West Bay isle in November 1528 carried the ultimate in maritime technology when they set sail from Cuba ten months earlier. Compasses and compass cards, knotted log ropes and lead lines, astrolabes and cross-staffs and nocturnals—all helped them determine the location, speed, and direction of their ships at sea.[7] The ships themselves were also advanced: five caravels—slim hulled, shallow keeled, and easy to maneuver in coastal waters—and one brigantine, smaller and outfitted with benches for oarsmen. Even their commander, Pánfilo de Narváez, had the highest of qualifications: he was a conquistador ordered by Charles V, emperor of Spain, to "discover and conquer and populate the land from the Rio de las Palmas to the island of *Florida*"—thirteen hundred miles of territory, starting well south of the Rio Grande, curving upward and across the Gulf, and arching down the Florida peninsula.[8] Equipped with the most modern of navigation devices and blessed with the approval of Europe's most powerful ruler, the small fleet should have had no problems—except for the immensity of the land they approached and the inadequacy of their pilot.

Of all persons besides the captain of a craft, the pilot was most important. Trained from youth to observe the shape of shorelines, educated in the location of landmarks, indoctrinated in the complexities of dead reckoning and celestial navigation, these men were the mainstay of every fleet. Upon their skills rested the safety of the crew and the success of their commander's enterprise. Pilots were expected to know how to maneuver across shallow bays, twist through narrow straits, and avoid becalmed seas. At the very least, they were to be familiar with the water bodies upon which they sailed. So, when Pánfilo de Narváez was unable to find a pilot in Spain knowledgeable about the region he was to conquer, he hired one in Cuba by the name of Diego Miruelo.[9]

Miruelo was as knowledgeable about the southern boundary of Narváez's promised territory as any one navigator could be; he had been to the Rio de las Palmas twice. The northernmost reach of Spanish power on the American mainland by 1527, the river had been included in the map Hernán Cortés sent Charles V five years earlier. According to this and another drawn in 1519, of the entire northern Gulf coast Narváez had been granted, there were only four locations worth noting: that of the Mississippi River, the tip of Florida, and two minor bays in between. No one Narváez knew had reconnoitered any of those areas; only Miruelo claimed to know one, Rio de las Palmas. So it was he the commander hired.[10]

To his everlasting dismay. The expedition had not even left the Cuban mainland before Miruelo ran the fleet aground. Once refloated, they continued west only to be blown off course again, this time by storms. By the time the lookout spotted land on Easter Sunday 1528, the ships had been at sea long enough to reach the bay of Rio de las Palmas. And so they had, Miruelo assured his commander. Within a week the pilot realized he was wrong, but he had sailed away by then, leaving his commander, over three hundred men, some women, and a few horses on the shores of the Florida peninsula.[11]

Seven hundred miles and eight months later, fewer than eighty crawled onto that West Bay islet. Remnants of Narváez's desperate attempt to cross the Gulf on makeshift barges, some had landed on Mustang Island farther south; others had survived long enough to come ashore at Matagorda Bay. But those were all dead by late November 1528. The few who made it to Isla de Malhado counted their blessings—and started scrounging for roots and rats to feed the few islanders who would tolerate their presence. Subject to orders, the survivors shelled oysters, roasted seedpods, and carried burdens. Subservient and obedient, they loaded dugouts for onshore moves, set up campsites on shell dumps, and trailed inland when their captors foraged for berries. For the island people had indeed become captors, their original curiosity about the intruders soon turning into scorn. Starvation, cold, and disease killed many of the castaways in the following five years; so did the natives they had disturbed.[12]

It was for these reasons, among others, that in 1533 four remaining Spaniards determined to escape their guards and journey to New Spain settlements in the south. They finally succeeded, although they ended up much farther west than they had intended. But their experiences stayed vivid, and it was in *The Joint Report* and the *Relación of 1542* that their leader, Álvar Núñez Cabeza de Vaca, immortalized the peoples they had encountered: the Han and Capoques who inhabited the Malhado isle where the castaways first landed; the Charrucos and the Mariames who lived onshore; the Avarares who camped along the Nueces River; and the People of the Figs, who poled dugouts along the back side of Padre Island. Cabeza de Vaca and his companions, Andrés Dorantes, Alonso del Castillo, and Estevanico, owed their lives to the people of the Texas coast. They returned the favor by passing out of their existence as soon as they could.[13]

But their image—bearded scavengers of various hues who spoke unintelligibly and knew little about life—remained vivid in the collective memory of the Indians on the Gulf, whose descendants watched warily when similar beings struggled ashore years later.[14] By the late 1600s, coastal natives had organized more thoroughly and their common tongue had given their culture a name—Cla-ma-co-ah, or Karankawa. They continued to live along the shores and islands of the lagoons as they had earlier, and they continued their inland excursions in small groups. But the "wandering tribes" had developed far beyond the loosely knit bands Cabeza de Vaca described generations earlier. They had become an organized culture with chiefs for peace and for war. And it was war the Karankawas feared when René-Robert Cavelier, sieur de La Salle, landed his shallop on Matagorda Island in 1685.[15]

The incursion that January day had not been a surprise. For months, alerted by the Louisiana Akokisa of three large vessels sailing west, Gulf peoples had watched the interlopers make their way along the coast. La Salle's actions—his clumsy sandbar landing off the Sabine river, his futile outreach to Karankawas in Galveston Bay, his careless abandonment of a ship near Matagorda Bay—all confirmed native unease. But it was the Frenchmen's despoiling of the barrier isles and the coastline that confirmed their most desperate fears. Everything that made their estuaries viable—sea grasses, mudflats, freshwater flows, marine organisms—was crushed under the thick boots of sailors as they waded ashore. The heavy boats the mariners beached furrowed shell banks and trenched crab hideaways. The cannonballs they shot panicked marsh birds; the sound vibrations shattered spawning pools. The grappling hooks they used to pull their shallops through the passes ravaged vegetation; the signal fires they lit scorched flats.[16] It was not until their commander ordered soldiers onto Matagorda Island, however, that islanders realized intrusion had become occupation.

Long convinced that only he could conquer Spain's northern Gulf coast and plunder its New Biscay mines, sieur de La Salle had persuaded royal, clerical, and private clients to support his expedition. It sounded good—a small flotilla easing into the Gulf of Mexico, careful to avoid the Armada de Barlovento; a military squadron deploying onto the shores of the rivière Mississippi, equipped to set up a home base; and a passenger hold teeming with civilians eager to grow their own crops.[17] The greatest selling point, however, was technical. Three years earlier, the explorer had sailed from the Illinois River to the mouth of the Mississippi. There, he had taken its measurements: about 27 degrees north of the equator and 24 degrees west of Quebec. With calculations based on maps available at the time, it was evident that when he landed his fleet on the "mouth of the river I intend to enter, quite at the far end of the Gulf's Bend," La Salle would be close enough to New Spain that he could attack its northern territory at will and still protect his colony's home base.[18]

The problem, however, was as technical as the sales pitch: La Salle's calculations were wrong. The seven-inch astrolabe he had used at the mouth of the Mississippi in 1682 was off by at least two degrees. Consequently, the location he gave for that site's equatorial distance was inaccurate by as much as sixty nautical miles. In addition, a ship's clock that could accurately measure time in relation to a given place had not yet been invented, so La Salle's longitude measurements were skewed as well. And the explorer knew this. First, he hedged on latitude, "about 27 degrees," and then on longitude when confronted with a map that showed another river, the Escondido, where he had earlier positioned the Mississippi. Suddenly, longitude changed, and with it, any uncertainty. "If all the maps are not worthless," La Salle wrote triumphantly, "the mouth of the *fleuve* Colbert is near Mexico . . . this Escondido [river] is assuredly the Mississippi." That, then, is where he and his squadron thought they were heading when they sailed out of French Saint-Domingue in late 1684. Where they were actually going was to the central Gulf and the barrier isles of the Karankawa Indians, for that is where the Escondido River flowed.[19]

The mistake was as damaging to the French as to the islands and the natives they invaded. La Salle had already lost one ship and its provisions before leaving Saint-Domingue. His miscalculations and a strong longshore drift put him far west of the Mississippi before he even sighted land in December, and the subsequent separation of his only warship left him coasting unhappily down St. Joseph and Mustang Islands for two weeks before finally reuniting off Matagorda Bay.[20] Within thirteen months, that

A nocturnal, a navigational device used by seventeenth-century pilots at night to determine approximate distance east or west. This was onboard La Salle's *La Belle* when it ran aground in Matagorda Bay. Courtesy of Texas Maritime Museum and Corpus Christi Museum of Science and History.

Above, lead line base, used to determine water depths; *below*, pocket-sized sundial, used to determine distance north or south. Both instruments were part of *La Belle*'s navigational equipment. Courtesy of Corpus Christi Museum of Science and History.

warship had sailed away permanently, La Salle's storage ship had grounded off Matagorda Island, and the only seagoing vessel left had been demolished following a February norther.[21] Almost as discouraging were the actions of the commander himself: continual absences in times of stress, habitual tirades at the expense of his underlings, obdurate refusal to take advice. Even his regret that, during the sixteen months of his occupation of Matagorda Bay, half of his company had died, was so phrased that the blame lay on the dead themselves.[22]

But it was not his compatriots who most regretted sieur de La Salle's appropriation of the islands of the coast; it was the people who inhabited them. For decades the Karankawas had dwelled on Matagorda, St. Joseph, Mustang, and Padre Islands. For decades they had harvested them in wintertime and camped near them in the summer. For decades they had defended them against Caddos to the east and Jumanos in the southwest. Now, mere months of intruder occupation had seen their streams contaminated, their estuaries crushed, their lagoons strewn with wreckage, and their passes blocked. Unfettered pigs scavenged their food stashes; boars wallowed out their creeks. Deer carcasses rotted on nearby plains, and bloated bodies polluted inland streams. Staked claims were ignored; attempts to converse were met with gunfire. Canoes disappeared, as did the occasional woman.[23]

There was no doubt their universe was at odds. Islanders knew it needed to be set right, but just as certainly, they knew that it could be done only in La Salle's absence. The man, whose anger was so scathing he drove his own men to desertion, was just as formidable to his native opponents. Consequently, they waited for him to leave his post, as he had already done twice before. Finally, he did so, not quite two years to the day after his troops first debarked onto Matagorda Island. On January 12, 1687, he took over a dozen men north, leaving the rest inland with a priest, a doctor, two mothers, and several children. He never returned; nor would they have greeted him if he had. It took time for the island Indians to hear of the commander's murder three months later, and even more time to confirm it. But in early 1689, knowing he would never bother them again, the Karankawas of Matagorda Bay slaughtered every man and woman in the interlopers' settlement, keeping only five youngsters alive. They even killed the pigs.[24]

But the universe did not right itself. News of French invaders had renewed interest among the descendants of the Spanish emperor who had claimed the area in 1526. Consequently, more intruders followed: Captain Pedro de Yriarte, whose ships bypassed the French post in 1687; Alonso de León, who found the settlement in ruins and retrieved some of the children from their captors two years later; and Domingo Terán de los Ríos, who recovered two more while exploring Texas in 1691.[25] Following on the military's heels were the Franciscans, who saw the people of the coast as neophytes, potential converts to the Spanish church. They set their first mission squarely upon the spot La Salle had commandeered thirty-five years earlier. To little surprise, the outreach was unsuccessful, and an attempt by presidio captain Domingo Ramón to massacre the islanders only intensified their hatred.[26]

The influx of Europeans was not to be stymied, however, any more than the determination of the priests. The mission was moved and moved again as padres worked to subject their charges to Christian life. The key, they felt, was food. "If the Indians . . . do not have enough to eat in their mission," one wrote, "they will wander about doing damage." Another wrote, "Solely to bind [to the mission] this portion of [a group of Karankawas] I gave them gifts . . . of beef, corn [and] tobacco." A third wrote, "We gave them what we could . . . ears of maize, maize flour [and] tobacco." In return, "they received us very well."[27]

Indeed they did, for in their struggle to maintain seasonal sojourns off the islands, the natives were using the missions as inland havens. The cattle there were not too different from the diminishing bison herds on the plains, and they were far easier to catch. Shelters were more substantial, and the skills their women were taught—weaving and spinning—were interesting. By 1791, the Karankawas' only request to the priests was "that you put a mission here on the coast . . . for us. We will gather in it . . . and we will bring all the heathen that are on the coast from the mouth of the Nueces to the Colorado River . . . then you can say the whole coast is yours." And they were as good as their word. Nuestra Señora del Refugio, the only Karankawa mission ever boasting a population over one hundred, soon nestled between Blanco and Medio Creeks, only miles away from Copano Bay. Established in 1793, it enabled the islanders to blend expediency with tradition, at least for a while. "They can be considered

. . . half-tamed," one chronicler noted in the 1820s, "com[ing] only occasionally to visit the presidio . . . hunting and more often fishing for their needs." Almost enviously he called them "mariners on the bays."[28] As did Colonel Juan Nepomuceno Almonte in his "Statistical Report" to the vice president of Mexico years later. "The Karankawas . . . could be advantageously used in the navy," he acknowledged. "They are practically reared in the water . . . excellent fishermen and good swimmers."[29] And so they were, as long as they stayed by their shores. But time, the islanders' old ally, was turning against them.

By the mid-1800s, the Franciscans were gone, the Spanish had decamped, and Almonte's government had been driven from the Coastal Bend. Just a few years later, the Karankawas were as well. Forced south of the Lavaca River by Stephen F. Austin, they were hassled down the Nueces and then driven west along the Rio Grande. Finally, attacked by rancheros from both sides of the border, the displaced islanders were annihilated in 1858. Ignominious and bereft, they died far from the windswept waterways that had made them "the best pilots in Matagorda and Aransaso Bays."[30]

PART II. EXPLOSIVE DAYS

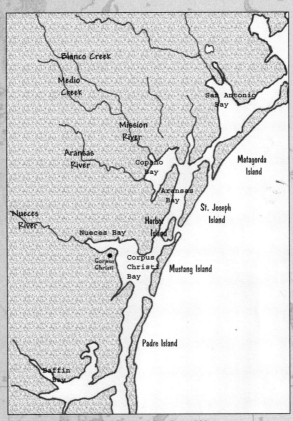

Copano and Corpus Christi Bays, mid- to late 1800s.

CHAPTER 3

Horse Marines

Whatever potential Colonel Almonte may have seen in Karankawan mariners, the Indians' demise was inevitable. Technology had overtaken the handcrafted dugouts they had maneuvered among the barrier islands of the Coastal Bend. Schooners now rode the waves washing the beaches; they and smaller skiffs coursed the estuaries of the Texas mainland. Such advancements were no surprise. Thoroughly convinced of the power of wind and finally capable of harnessing it, ship designers had progressed from caravels to more streamlined vessels. Long, slender, shallow drafted, and only two masted, schooners were light enough to sail

Schooner in Corpus Christi Bay, around 1900. In design it remained almost identical to those of the early 1800s. Courtesy of Jim Moloney.

into lagoons and flexible enough to maneuver through most inlets. Their smallish hulls, narrow at each end, held another advantage; they provided enough room for only a limited cargo, generally the pirated kind.[1] For, besides confronting rebels against his government, now led by General Antonio López de Santa Anna, Almonte was facing another problem in 1835: prolific smuggling off the Texas coast.

It was to be expected. The nation of Mexico, although independent from Spain for over ten years, had continued the mother country's suspicion of colonial seaborne trade. Stridently harangued by prominent empresarios, the central government had reluctantly authorized a port on Galveston Island in 1825. But tariff regulations, so onerous that even its own trade commissioner balked, followed two years later, and additional restrictions, like that demanding that clearance papers be notarized at only one location, compounded colonists' frustration. Their ad hoc blockade, which prevented the landing of supplies for Mexican garrisons off Galveston Bay, finally forced the seated government to remove trade barriers in 1832, and Texans rejoiced at the validation of the "Turtle Bayou Resolutions."[2]

But their triumph was short-lived. Incensed that smuggling along bays and rivers had increased even as tariffs decreased, Santa Anna set up additional customs houses, increased tariffs yet again, and sent his premier schooner, the cannon-laden *Moctezuma*, to patrol the coast. But it was not until the generalissimo hired vessels to supply troops garrisoned near Copano Bay that he felt the full brunt of Texan guile.[3]

Part of the coastal system stretching south from East Bay and harboring its own tiny inlet, Copano Bay was an ancient basin that had reflooded into a nurseryland of life. Fed by the Aransas River to the south and the Mission in the west, its bottom hosted embryonic oysters and sea grasses; its lower waters teemed with mud creatures and microscopic plants; and its shores harbored tidal flats and marshes. Hugged by peninsulas through which its waters flowed into Aransas Bay, then through the pass at St. Joseph Island and finally into the sea, it claimed the name of a tribe still there in 1700.

Like the rest of the Karankawas, these Copanos were seasonally nomadic, with "huts made of hide that they fold like tents and carry with them," noted French explorer Jean Bérenger, who came upon the bay in 1720. But they had a small permanent village "of about a dozen large, quite round huts . . . where they put the supply . . . of fish that they dry without salt."

Live oaks along Copano Bay. Courtesy of Jim Moloney.

Curious about their worship—"I could not find out anything about their religion,"—their sustenance—"They eat . . . fish-half raw . . . and oysters"—and their prognostications—"The number of [a snake's] rattles . . . has efficacy for the observations they have made"—Bérenger was most taken with the natives' innovations. Logs burned into pirogues, bones sharpened into scrapers, coral whips twisted into fishing lines, mulberry bark fashioned into rope: all made entry into his journal. But even more astounding to the captain were the formidable marine deposits amid which the Copanos lived. "I went five leagues into the bay to reconnoiter the lay of this island," he noted. "The mainland was still three or four leagues distant . . . [but] an oyster reef [kept] me . . . from there even with my launch."[4]

His departure days later was no easier, a submerged shoal holding Bérenger's ship aground until the "currents bearing out to sea" tugged him away. Time and disease took the natives, and gradually the region returned to nature. Sporadic suggestions that Aransas and Copano Bays become part of an armed defense line against intruders went largely unheeded, and by the early 1830s, the area had reasserted primal life. Currents layered more silt along peninsulas as oak trees spread atop them.

Dunes lurched farther westward on St. Joseph, and new passes cut between it and Mustang Island. Oyster reefs grew higher and wider, extending thickly along the bay bottom. By the time an immigrant-laden schooner tried to access Copano in 1833, the sandbar between it and Aransas Bay was so dense it took three days for the pilot to cross. Even within the cove, the colonists had to position a wagon into the water, load it "from the boat," one recalled, "and then haul it to shore . . . as we could not approach within 400 yards of the beach."[5]

Despite the difficulties, this was the inlet Santa Anna had been using to offload munitions during the Texas rebellion, and it was here that additional supplies were being sent, even as he negotiated surrender terms after his San Jacinto defeat. As worried as General Sam Houston that Mexican troops garrisoned inland would remain a threat, volunteer riders, under the leadership of Major Isaac Burton, began to patrol Copano's coastline, on the lookout for advancing ships. The approach of one on June 2, 1836, confirmed their suspicions, as did its flag of Mexico, raised after distress signals from shore duped the captain into showing his colors. Leaving the schooner and rowing carefully to the beach, he and five crew members were taken by Burton's "horse marines," while others piloted the sloop back, overcame the remaining crew, and confiscated their cargo. Repeating the maneuver successfully over the next months seemed to safeguard the new republic from future coastal intrusion, as did the creation of a Texas navy. But it was not until both governments—Texas and Mexico—signed a treaty seven years later that Aransas and Copano Bays and the barrier isle that defended them seem secure.[6]

But the gods of war were fickle at best. Not thirty-six months afterward, troop ships approached St. Joseph Island once again. Their ultimate destination this time, however, lay sixteen miles farther south. There, almost half of the US Army would set up camp along one of the most controversial inlets of Texas: Corpus Christi Bay.[7]

Even its name evoked question. Like most of the coastal entities explored by Spain and France, the large bay, with sister bay and estuaries, had been given many titles. Its primary river, the Nueces, was the original Rio Escondido targeted by La Salle in 1684. A barrier isle fronting the bay had been christened Isla Blanca by Alonso Álvarez de Pineda in 1519 and San Carlos de los Malaquittas by Diego Ortiz in 1766, becoming Padre Island only in the early 1800s.[8] The great bay itself—extending east from Nueces Bay in an almost completely round, 152-square-mile circle—was included in Pineda's original chart. But it was not until explorer Joaquin

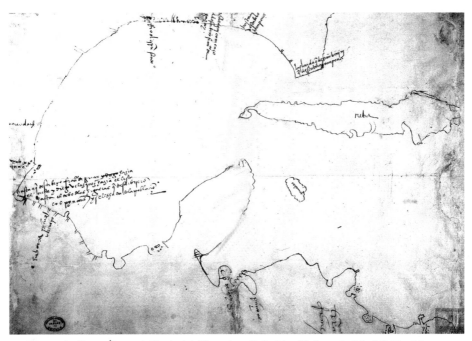

The map by Alonso Álvarez de Pineda. Isla Blanca (now Padre Island) is far to the left. Full title of the drawing is *Dibujo de la costa del golfo de México desde la peninsula de Florida hasta Nombre de Dios*. Courtesy of Archivo General de Indias, MP-MEXICO, 5.

de Orobio y Basterra christened it in 1746 that it had a label: San Miguel Arcangel. It was cartographer and surveyor Colonel Don Diego Ortiz Parrilla, however, who placed it on his map as "la Bahia nombrada Corpus Christi" in a report to Governor José de Escandón twenty years later. By 1835, the bay had a name both official and awe inspiring.[9]

It was more than nomenclature, however, that made the area intriguing. The bay shore itself was unusual. Although its coastline featured sand extending in a wide strip northward to Matagorda, it backed onto a bluff of blackland clay that rose over fifty feet above the beach. Its vegetation also varied. Where live oak were so prolific they adorned the peninsula and isles abutting Copano Bay, greenery along Corpus Christi Bay consisted of creosote bushes and mescal bean shrubs. Its marine deposits differed as well. Any crossing from one side of the Aransas inlet to the other required boats, sloops, or pirogues. But an ancient oyster reef within Nueces and Corpus Christi Bays had grown to such a height it approached water level—and provided an occasional but effective walkway between the two.[10]

The areas' protective barriers were also unique. Although peninsulas alternated with isles to shelter Copano and San Antonio Bays upcoast, it took a single island, Mustang, in conjunction with the tip of St. Joseph to the north and the edge of Padre to the south, to shield Corpus Christi Bay from the sea.[11]

Possibly the most notorious feature of the bay, though, was its inhabitants. Those noticed by Cabeza de Vaca on his trek southward, the Avarares and the Fig People, had long since vanished with the Karankawas. With the exception of occasional wanderers, few Indians lived there anymore—but miscreants did "and lawless persons to whom war was prosperity and satisfaction," commented a contemporary. They dwelled in "a small . . . hamlet . . . containing probably less than a hundred souls," consisting of "some twenty or thirty houses, and just two bars."[12] Hidden by barrier islands to the east, accessed with difficulty from the west, and peopled by degenerates on its bluff, the settlement was a smugglers' paradise, led by its own potentate.

On the run from Illinois, Henry L. Kinney had arrived at Aransas Bay years earlier, then migrated south, where he created his own domain—Kinney's Rancho—along the western edge of Corpus Christi Bay. "He seemed above all laws, save those he established for himself," a chronicler observed. "Alternately the friend and foe of Mexicans, Texans, Americans, and Indians, sometimes defying them and meeting them with force and sometimes bribing and wheedling them," Kinney ruled his marauders with expertise.[13] Under his direction, they ran tobacco to Mexico and wool to the States, paying neither tariff nor duty in the transfer. Confronted with a competitor farther south, they arranged for his murder. Attacked by Comanches, they slaughtered nearly all. Threatened by outsiders, they armed the bluff. Regarded suspiciously because of their locale, they rechristened their village with the name of the bay.[14]

But they could neither whitewash their reputation nor relocate their site. And it was Corpus Christi Bay that drew American troops southward that early August dawn of 1845. Of all spots along the coast, it was upon its edge, south of the Nueces and north of the Rio Grande, that General Zachary Taylor of the US Army of Observation hoped to instigate a war.

CHAPTER 4

Sad Havoc

Not by his own choice, however—at least not at first. The forces that sent Taylor and most of his Third Infantry from Louisiana, down the Mississippi to New Orleans, and across the Gulf had been unleashed years earlier.

Signing treaties at Velasco, not far from his defeat at San Jacinto, seemed to have meant comparatively little to Generalissimo Antonio López de

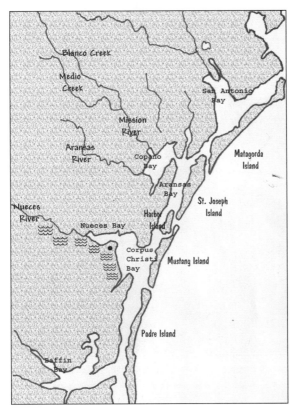

General Zachary Taylor's army along Corpus Christi Bay. Based roughly on Daniel Whiting, *A Soldier's Life*.

Santa Anna in May 1836. His capture at the hands of the Texas army meant far more, as did his desire to return to Mexico City. So agreeing to send his troops south of the Rio Grande and acknowledging that river as Mexico's northern border may have come easily. The choice of rivers, however, was portentous. Up to 1824, the Rio Grande had always defined the boundary of Coahuila y Texas on Mexican maps. But the government had redrawn its possessions and by 1835, the Great River lay in the territory of Tamaulipas, and Corpus Christi Bay's Nueces River signaled Texas's southern border—a loss of hundreds of square miles. By reestablishing the Rio Grande as their southernmost boundary, the drafters of the Velasco Treaties were not only restoring lost lands but also absorbing all the territory along the Great River north through New Mexico and beyond. Had Velasco gone into effect, the size of Texas would have more than doubled immediately.[1]

That the accords did not go into effect was due more to the faithlessness of both countries than to the righteousness of the agreements. What remained, however, was their effect: with the Nueces River as boundary, Mexico kept its territories of Coahuila, Tamaulipas, and Nuevo Mexico.

Daniel Whiting's lithograph of Taylor's troop encampment on the banks of Corpus Christi Bay. *Bird's Eye View of the Camp of the Army of Occupation, near Corpus Christi, Texas (from the North).* Courtesy of Corpus Christi Public Libraries.

Sad Havoc

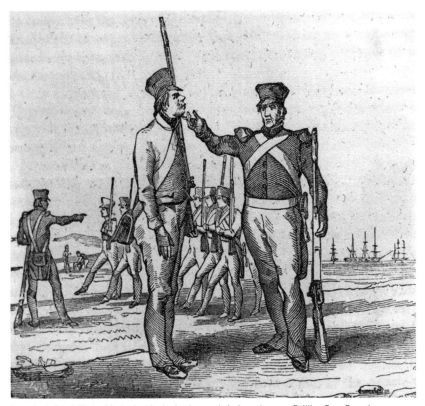

Recruits being drilled with the bay in the background. Artist unknown, *Drilling Raw Recruits*, engraving by William Croome, from John Frost, *Pictorial History of Mexico and the Mexican War*. University of Texas at Arlington Libraries Special Collections, Gift of Jenkins Garrett.

With the Rio Grande as boundary, most of these lands became Texas, and thus pivotal in America's Manifest Destiny. The nation's march from sea to sea that had begun with acquisitions from France, Great Britain, and Spain would accelerate when it annexed the Republic.

But Mexico refused to cooperate. Not only was it reinforcing its claim to the Rio Grande militarily, but it was ignoring an American offer to buy its western provinces. Any attempt by the United States to extend its boundaries south of the Nueces by annexing a Rio Grande–bordered Texas was asking for trouble. Why, then, in the spring of 1845 did Secretary of War William L. Marcy order Zachary Taylor to take the forces "under his command . . . and approach as near the western boundary of Texas (the Rio Grande) as circumstances will permit?"[2]

Officially, it was to guarantee the soon-to-be-annexed new state "defense and protection from foreign invasion." Unofficially, as Ulysses S. Grant put it years later, "We were sent to provoke a fight."[3]

"But it was essential," the later president added, "that Mexico should commence it. It was very doubtful that Congress would declare war" over land itself, even if the territory in question was half again the size of Texas. "But if Mexico should attack our troops," ah! then "the Executive could . . . prosecute the contest with vigor." And so, obeying orders despite personal reservations, Taylor sailed his infantry and marched his dragoons across the Nueces River onto Mexican-claimed Corpus Christi Bay.[4]

Or rather, his dragoons crossed the river. His infantry—many of them—once they were moored on St. Joseph Isle, jumped the sides of the *Alabama* and waded fully clothed through surf onto shore. They knew, once they moved farther south to Corpus Christi, that their very presence should goad Mexico into war. But the sheer excitement of the venture, the coming together of troops for the first time since the War of 1812, and the exhilaration of making land after days at sea overrode any qualms. Even the passage across the Gulf had been breathtaking for some, embarking on vessels powered by steam-propelled paddle wheels rather than sails. Schooners still harbored in ports and slipped between barrier isles, but it was narrow-hulled steamships that carried many of Taylor's men to St. Joseph Island, and smaller steamboats that lightered them across the estuaries to the Bay of Corpus Christi.[5]

And there they camped, early arrivals fanning out below the bluff, Colonel David Emanuel Twiggs's dragoons spreading farther west along the Nueces, some artillery resting at the juncture of the bays, and remaining infantry tenting farther down the beach.[6] By October up to three thousand men were mustering, marching, drilling, and parading on the shoreline, a most appropriate "esplanade," one officer remarked, on which "to . . . drill . . . men . . . for all they were capable of."[7] Drills—and target practice— seemed all they were capable of at that moment, with so little reaction having come from the south. "Every-thing quiet on the Mexican frontier," one officer wrote his wife, "and not the slightest prospect of any collision."[8]

But where the enemy dawdled, Corpus Christi thrived—and Kinney with it. Within weeks after his arrival, Taylor was hiring headquarters from the overlord, taking his advice on Mexican relations, and utilizing his informants. The town itself expanded. "One can hardly realize," one soldier wrote in November, "that the Corpus Christi before us now is the settlement of scattered houses we saw upon our landing." Buildings grew overnight, he noted, as shacks rented out as commissaries, sheds emerged as brothels, huts reappeared as grog shops, and shanties became faro parlors. Population grew twenty times over as outsiders rushed in, "the majority of them . . . grocery keepers and gamblers, who have come to feed upon the army."[9]

The men responded as anticipated. By October there had already "been several disgraceful brawls and quarrels," a diarist wrote, "to say nothing of drunken frolics. . . . One captain has resigned . . . and two others . . . are still on trial for fighting over a low woman." Another commented on a subordinate arrested for fornication within his tent. "The Colonel . . . did not care how many liaisons his officers had outside the lines but within the encampment it was a very serious matter." Soon, "the village and house on the hill [bluff]" were declared off-limits to soldiers, and patrols were assigned to "apprehend . . . any enlisted man" found in that vicinity. Just weeks after the troops' initial arrival on Corpus Christi Bay, the stockade was filled to capacity and the first court martial convened. By March 1846, over two hundred soldiers had been arraigned before tribunals, mostly on charges of drunkenness and desertion. All but twenty-three were convicted.[10]

But for those less susceptible to sin, the bay and its surroundings were magnetic. Young Ulysses Grant, fresh from West Point and commander of the regimental mule train, was particularly attracted by its coastal plains. "It is just the kind of country," he wrote his fiancée, "that we have often spoken of in our most romantic conversations . . . a place where we could gallop over the prairies and start up deer and prairie birds and occasionally see droves of wild horses or an Indian wigwam." Its vastness awed him, as did its emptiness. "There was not at the time a [single] individual living between Corpus Christi and San Antonio—a distance computed at one hundred fifty miles—until about thirty miles of the latter."[11]

Grant also extolled the bayside climate as "delightful and very healthy . . . equal to that of any in the world." Others agreed. "It is probably one of the healthiest and pleasantest spots in the world," one news correspondent wrote. Another noted, "There was no sickness at all among the troops, the air was very fine and nights cool, with a good breeze night and day from the sea."[12]

The ultimate charm, however, was the provisions the prairies and inlets provided. "The disciples of Isaac Walton had rare sport in the bay and streams," one officer remarked, "and sportsmen a field for all kind of game." Captain William Henry described hunting deer on horseback, "with the excitement of shooting them at full run." Others simply sat, waiting for animals to come within range. The carcasses one set of officers brought back from an overland hunt included fifty-one geese, ten deer, four bittern, two cranes, eighteen ducks, three turkeys, and a seven-foot-long panther. "Fishing was just as profitable," Private Bern Upton bragged as he described the fifty to sixty bushels of fish his regiment captured most days. Both he and Captain Daniel Whiting remembered the winter's night when it got so cold that "masses of fish . . . oysters and turtles . . . were found stranded . . . on a reef." It took little effort at dawn to fill wagons and distribute the cold-stunned animals to cook fires throughout the camp.[13]

Little effort indeed was needed to stanch morning hunger, any more than it was expended to provide daily meat—too much was available. But abundant resources could not compensate for the sheer lassitude that soon permeated the camp along the bay. "What a pretty figure we cut here!" one commander complained about poorly drawn orders and sloppy maneuvers. Shoddy and criminally thin tents occasioned other outbursts, and high-ranking officers unexpectedly drew fire. "Even Colonel Twiggs could put the troops in line only 'after a fashion' of his own," one sniffed.[14] But it was abandonment of basic military procedures—safe and distant latrine locations, universal smallpox immunizations, well-built transportation vehicles—that began a slow but deadly contamination. The beach became grimed with feces, and the bay thickened with effluent. The offshore explosion of "an old hulk" of a steamship rendered waters putrid—and damned. "Eight killed and seventeen wounded," a chronicler wrote, "the poor, mangled fellows lay clinging to pieces of the wreck."[15] Regimental hospitals multiplied, and jaundice, diarrhea, and toothaches became as integral to the men as the sands they were fouling.[16]

Inexorably, the bay and its environs took revenge. "You miss the palmetto and pine," one soldier lamented. "It is said there is not a pine tree in this part of the country," another complained. "Were we where we could get trees, logs, boards, or anything of that matter, it would be a different matter." "We were forced to . . . plant chaparral to the north of our tents to break the wind," one officer reported, as well as build "embankments . . . [with] casks." Water was rare. Grant's observation that "the country does not abound in fresh water" was an understatement of magnitude. Water that existed was brackish and contaminated, "from which many have died and all have suffered," a diarist noted.[17] What did flourish, along the shores and tidal flats and within the lagoons and beaches, was disturbing. "We found the place infested with rattlesnakes," Captain Whiting related. "I was awakened one night by the rattle of one in my tent. . . . We frequently found scorpions in our clothing and boots." "There are millions of flies here," Quartermaster Napoleon Dana observed. "They won't let us sleep in the daytime, [they] cover the walls of tents at night, fly into our noses and mouths, get into our eatables. . . . This is a mighty thriving country for cockroaches too." Grit and dirt thrived as well. "It is almost impossible to keep clean, although the bay is just behind our tents where we can bathe whenever we wish. If I sit down to read or to write, I get my face, hands, eyes, papers all full of sand, sticks, etc." It was a last complaint, however, that was most revealing: "There is [this] constant wind blowing from the sea."[18]

For the bay had thrust northers onto its interlopers. An early one blew in on August 24 with such force it took "your breath away," Captain Henry remembered, making "you sit bold upright in your chair, feet on the rim, as if your life depended on it." More followed, contrasted with sudden spurts of warmth. "The change in temperature . . . is incredible," one officer recalled. "The thermometer will fall forty degrees in a few hours," another added, "and from having been burned by the sun, you are frozen by the cold air." Deeper winter winds tore even harder at the troops, sweeping in "gales, day and night," moaned Lieutenant George Meade, and preventing "you from getting your tent comfortable." The hastily thrown-up shelters of brush and kegs did not prevent drinking water from freezing, tents from leaking, or men from falling ill. As of late November, one-third of Taylor's troops were on the sick list, 10 percent of the officers were bedridden, and "the whole Army . . . might be considered a vast hospital," surgeon John B. Porter recorded. "Dysentery and catarrhal fever have made sad havoc among the troops."[19]

By the time General Taylor marched his men south to the Rio Grande—it was becoming "necessary for the 'invaders' to approach" closer to the border if hostilities were to begin, Grant noted dryly—most of his men were ready to leave the region. "Send me back to the United States," one cried, "but Stop! I forgot Texas belongs to Uncle Sam—and a mighty poor bargain he made, I think."[20]

As the Third Brigade—the first to arrive, the last to leave—moved away, Captain Henry glanced back. "Corpus Christi looked perfectly deserted," he wrote, "like desolation itself." As well it might; most original inhabitants were moving south with Taylor. "But," he added, "the bright waters of the bay looked as sweetly as ever."[21]

CHAPTER 5

Harbingers

As well they might, that bright March morning, the bay's greatest transgressors were leaving. But as the Karankawas had discovered earlier, less tumultuous times would never return. Coastal Bend waterways—Corpus Christi, Nueces, and Copano Bays, Laguna Madre and Aransas Bay channels, and North Padre, Mustang, and St. Joseph Islands—had branded themselves indelibly on the military mind. They demanded attention.

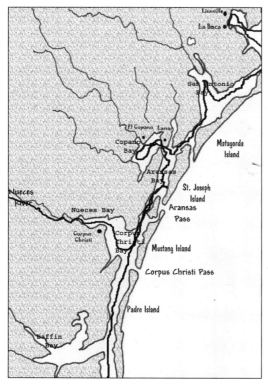

US Army reconnaissance teams, 1845–1846. Based roughly on *General Taylor's Life, Battles and Correspondence* and George Meade, *The Life and Letters of George Gordon Meade.*

General Taylor recognized this, even before he received this order from Secretary of War William Marcy: "It is extremely desirable that the seacoast, or at least that part of it which will be likely to be visited by our vessels . . . should be better known than it now is." In late September, he had sent a reconnaissance team up the Nueces River to ascertain the difficulties "through which his army would have to operate, in the event of an advance movement." Two months later another squad headed south down the Laguna Madre on a ten-day expedition, producing drawings and reports "of a most important nature." Early the next year, the general ordered a thorough exploration of Aransas Bay, this group to sail "up the coast along the inside passage . . . and visit the little towns of Copano, Lamar, La Baca, Linnville, and Matagorda, in succession."[1] Officially a marine expedition, traveling in open boats—probably the small schooner-rigged flat-bottomed vessels they had used navigating the Nueces—Taylor's engineers left St. Joseph Island on January 26. They spent the next month "knocking about the bays," one wrote, "making surveys . . . visiting towns, and places where towns are to be." Armed with sextants for measuring latitude and the tables in *American Practical Navigation* for longitude, they were far ahead of most nineteenth-century counterparts in determining location. Moreover, the new technique of using astronomical observations as a base for position lines on a chart helped them verify sites as they moved along the channels. But perhaps the most welcomed of the devices available for the topographers were the new logs that mechanically recorded distance and converted it to nautical miles. Mechanical depth sounders were not as common, so estimating the amount of room a vessel had to have in order to cross the many bars and shoals within the coastal waters still depended on bottom-trolling lead lines, as well as local wind and current maps.[2]

Adequately equipped but bedeviled by winter northers and rain, the boaters traveled the length of St. Joseph and Matagorda Islands, crossed into Copano, made landings in Aransas and Matagorda Bays, and finally came to rest on the tiny outcropping connecting the mainland with Matagorda Peninsula. Approximately one hundred miles south of the port of Galveston, they reveled in Matagorda, "one of the old settlements of the country," and sent letters home "through the Texas post-office." By February 24, they were back on St. Joseph's Island, planning to return to Taylor's encampment on one of the little steamers the army had requisitioned.[3]

Harbingers

Steamboat being offloaded at a river port. Courtesy of Texas Maritime Museum.

That the cartographers had reliable transportation back to Corpus Christi Bay was another indication of military might. "A movement of this kind brings into bold relief our grand system of internal navigation," Captain Henry remarked as he chronicled the landing of the Fifth Infantry in Texas. "Detroit was their starting-point; thence across to the Ohio River by canal; down the Ohio and Mississippi in steam-boats to New Orleans, and by the steam-ship *Alabama* to Aransas Bay." No less impressive were the designs of the vessels employed: steamboats paddled inland streams with flat bottoms and wide hulls, while steamships with deeper and narrower hulls traversed the seas. Thus the *Creole* joined the *Alabama* carrying troopers across the Gulf to St. Joseph Island, where the more river-worthy *Undine* waited to convey them to Corpus Christi Bay. But the army was parsimonious as well as practical; as a result, many troopers found themselves still in schooners, like Ulysses Grant. "Ocean steamers were not then common," he reminisced years later, "and [our] passage was made in sailing vessels."[4] But it was not getting to St. Joseph Island that became the problem for Zachary Taylor and his troops; it was getting from there to Corpus Christi Bay. Herein lay one of the greatest drawbacks of the waterways of the coast: they silted.

Of course they did. The whole foundation of the islands and the channels they fostered depended on the sands that flowed onto them. Particles dropped alongside passes, sediment sifted onto deltas, particulates squeezed into bottom-based cores—all were part of barriers' life. They grew by currents of wind and water and diminished by the same agents. But the process was continual, and what was normal for the Coastal Bend—impervious sandbars, shallow bay bottoms, extensive shoal enlargements—was anathema to the military. Consequently, despite the position of two passes convenient to Corpus Christi Bay—Corpus Christi Pass on the south end of Mustang Island and Aransas Pass at its north—only the second one had enough depth to allow steamships and schooners to approach.[5] Once through Aransas Pass, schooners could anchor near Shell Island, as Grant's *Suviah* did, and painstakingly offload men, baggage, artillery, and garrison equipment onto the decks of smaller steam packets. Steamships, however, generally displaced too much water even to cross the eight-foot-high bar, so they moored on the seaward side of St. Joseph Island, from which they lightered their cargo and men onto steamboats.[6]

The whole process was extraordinarily time consuming in its own right. That the river packets oftentimes themselves displaced too much water to cross inland was almost unbearable. Grant noted the problem when his steamer, "as small as it was, had to be dragged over the bottom" to Corpus Christi Bay. But the situation became monumental when it happened to the army's commander. "On the morning of [July] 29th," Colonel Hitchcock wrote, "General Taylor determined to take two companies aboard the lighter *Undine* and attempt to pass down the bay of Aransas into that of Corpus Christi, between the island and the mainland. The difficulty was that the lighter drew more than 4 feet of water. . . . We . . . ran aground about five miles down the bay." Marooned for two nights, and "quite beside himself with anxiety, fatigue, and passion," Taylor finally secured enough smaller boats to transfer his men and baggage to the encampment. But he returned to St. Joseph Island, sequestering himself there another few weeks before belatedly making headquarters on Corpus Christi Bay.[7]

Coastal Bend waterways were not easy to traverse, especially for soldiers like Zachary Taylor and George Meade, who preferred "land expeditions . . . and being on horseback." Engineer Captain George McClelland was no less negative seven years later: "Government stores should never be sent to Corpus Christi Pass . . . when it is possible to send to Aransas." But the passes proved just as onerous for US sailors. Not quick on the scene—the army established a presence in South Texas within weeks after Taylor's departure—the navy still maintained a distinct interest in the barrier isles.[8] The soundings and configurations Taylor's topographers had documented in their 1846 expedition up the Aransas and Matagorda waters found their way into coastal surveys, and charts of the area were becoming available. Even more helpful were signal points placed on the islands to help navigators access the bays. Captain Monroe used the pole on the southern end of St. Joseph to clear Aransas Pass in 1833, and Lieutenant Craven sighted the one on the northern tip of Mustang to run the bar eighteen years later. The 1853 sailing directions by Lieutenant H. S. Stillage employed two posts: "Bring the signal poles on St. José Island in range, then stand up the Channel. . . . When nearly up to the point of Mustang Island, stand across . . . and run in 'til over the Bar." But it was the lighthouses the government installed—one next to Aransas Pass in 1857 and another on the bluff overlooking Corpus Christi Bay the same year—that revealed Washington's increasing concern about the Coastal Bend.[9] For despite their inaccessibility, the islands and bays had become integral to South Texas, and South Texas was planning to secede.

Not all by itself. By 1858 animosity toward the federal government had permeated most of Texas. The repercussions of the Dred Scott decision and the Kansas-Nebraska Act had already sparked movements to reinforce white supremacy in the state, and editorials excoriating "The Negro," "Negro Citizens," and "The Northern Democracy" found full acceptance in Nueces County. It was federal troops' withdrawal from the Rio Grande, however, and their subsequent delay in repulsing border marauders, that heightened South Texan disgust. Despite the eventual appointment of Colonel Robert E. Lee to military command, with explicit orders to "pursue [bandits] into Mexico, if necessary," regional faith in the United States was expiring.[10] Rumors of abolitionist-inspired arson hastened its demise, and the election of Republican Abraham Lincoln to the presidency destroyed it. By the time the state declared its loyalty to the Confederacy, the people on Corpus Christi Bay had already taken "possession of General Taylor's old camping grounds" and were stockpiling weapons for war.[11]

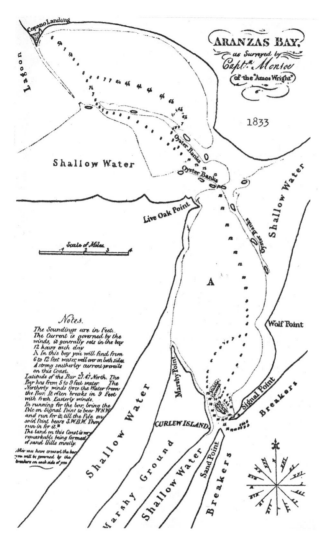

Captain Monroe's chart giving directions on getting through Aransas Pass. *Aranzas Bay, as Surveyed by Capt. Monroe of the Amos Wright, 1833.* Courtesy of Port Aransas Museum.

Nor were they the only ones. The "little towns" along the channels charted in 1846 had grown into sizable communities by 1861; El Copano now had three wharves, and La Baca (eventually known as Lavaca) was a significant port. Combined with emerging settlements like St. Mary's, downshore from El Copano, and the Mercer enclave on Mustang Island, the inhabitants of the barrier isles far outnumbered those left behind when

Taylor moved south. Their resistance to change was no less strong, however; just months after Texas seceded from the Union, they joined citizens from six coastal counties demanding the state Confederacy build "a fortification at Aransas Pass." Determined to defend Southern lifestyle, they even pledged personal funds to cover the expense, then requested a battery of artillery from military headquarters in San Antonio.[12] But times were already changing for them—at their very feet.

Or rather, in their very waterways. The long wharves extending into Copano from the shoreside of St. Mary's had destroyed bay-bottom habitats while being built, and their duplicates in Corpus Christi and Aransas had done the same. Dredge boats stopping the flow of the Nueces in 1861 had intensified bay salinity, and their buckets ravaged sea-grass meadows in nearby estuaries. Perhaps even deadlier than the mounds of mud scavenged by the dredgers and redumped at will was the sediment they loosened. Clouded waters blocked sunlight and stifled marine growth in a way that had never happened before.[13]

The murkiness dissipated in time, however, and bay-bottom colonies would rebuild. But changes were occurring on the barriers as well. Cattle brought by immigrants like Robert Mercer thrived on the rushes and sedges of island interiors. They were few at first, and their presence did little to threaten the animals already there, but their grazing did. Lush pepperwort, smartweed, and burhead vegetation that had flourished in shallow depressions grew sparser, as did some salt-tolerant shrubs. Settlers' horses, pigs, and chickens roamed freely, while indigenous birds began to disappear. The original 1840s assessment one riverman gave of the islands—"as green as a garden"—was beginning to go awry.[14]

But no injury inflicted by islanders or entrepreneurs would equal that done by war. It started mildly, mere rumors the Union was to blockade the Confederate Gulf. The strategy made sense. With the possible exception of a western territorial conquest by Henry Hopkins Sibley—seen as a pie-in-the-sky proposition[15] even at the time—the only ports easily available for British trade to the Confederacy by mid-1862 were along the Gulf of Mexico. Atlantic outlets like Cape Hatteras and St. Augustine had been stifled by the US Navy, barrier islands along the Carolinas already sheltered northern troops, and shipping centers from Fort Macon to Jacksonville were being quietly abandoned by commercial fleets. Ulysses Grant,

Taylor's young lieutenant from sixteen years earlier, had broken the river forts of Henry and Donelson, and Flag Officer D. G. Farragut took New Orleans in late April.[16]

If the South was to defend its territory, procure more ordnance, and supply food, uniforms, and medicine to its people, remaining Gulf ports had to stay open. Thus Lincoln and his strategists reasoned, and thus just months after Coastal Bend islanders petitioned their government for fortifications, a Union gunboat appeared off Aransas Pass.[17]

CHAPTER 6

Hellfire

But this was no standard Union gunboat. Pressed into service by the war, the USS *Arthur* had begun existence as a merchant schooner, square rigged on two main masts with a fore-and-aft sail on its mizzen and enough cargo space to provision the entire mercantile Gulf. By early 1862, however, it had been refitted: artillery carriages supporting Parrott guns replaced midship storage bins, and cases holding solid shot, shells,

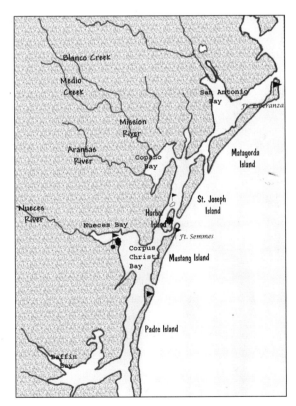

Fortifications and lighthouses along the Coastal Bend, 1861–1865. Based roughly on Norman Delaney, *The Maltby Brothers' Civil War*, and *Mustang Island Map of 1846*, University of Texas Marine Science Institute.

canisters, and grape ranged deck-side. Its commander had been refitted as well. Earlier a Great Lakes captain and later a transport skipper operating out of Brazos Santiago, John W. Kittredge officially became an acting volunteer lieutenant of the US Navy when the war started; unofficially, he was a tyro bristling for a fight.[1]

A fight that might never occur. Sedimentation had built the Confederacy's best bulwark: the sandbars on Coastal Bend passes were far too high and wide to allow schooners like the *Arthur* to enter. But anyone owning a screw-propeller scow, flat-bottomed boat, small steam packet, or sloop—and armed with a local's knowledge of reefs, mudflats, and water depths—could cruise the inner channels virtually at will. These blockade runners risked a deadly commerce, slipping along inlets like Aransas Pass or Cedar Bayou to drop off ammunition, foodstuffs, and supplies from Matamoros and Brownsville and uploading bales of cotton and wool to carry back. Never completely safe and well aware of the growing Union threat, such adventurers sustained the South—and infuriated Kittredge.[2]

But he bided his time and worked from the sea, methodically waylaying some vessels and confiscating cargo in January, gutting and abandoning others in February, and forcing the destruction of several in August. Intermittently, he goaded the islanders, sailing close enough to the Mercer settlement on Mustang to draw fire, then ravaging it and another, leaving the hamlets in shambles. Prideful, he challenged Confederate reports of his activities; vengeful, he spearheaded raids on homesteads and farms.[3]

Kittredge grew increasingly contemptuous in April when his guns shattered a rebel attempt to set a battery near Aransas Pass. (Resited on Shell Bank Island, the fortification stayed safely out of range but was of questionable impact.) Even the lieutenant's near capture that same month left him unfazed, the loss of two launches rationalized as part of his need "to escape . . . a greatly superior force."[4]

While he and the *Arthur* flitted off the coast, tension mounted on the land. Suspicions flared in March with rumors of currency manipulation and domestic terrorists, and a sedition law went into effect, condemning any who encouraged people "to favor the enemy." Despite initial rebel enthusiasm, the Confederacy had to institute conscription in mid-April and Texas governor Lubbock enforced it with martial law. Fears of invasion spiked, accompanying statewide cries to deal with dissenters "as traitors

Hellfire

deserve to be."[5] But it was the attempted desertion of draftees from the new Shell Bank fort in May that unleashed Coastal Bend fury. Incensed that the culprits' goal was Kittredge's ship, townspeople put the two survivors on trial, hanged them publicly, and coerced suspect residents to watch them die. The wave of distrust was so virulent that longtime citizens left, some fleeing the country. But the tension remained, settling like a grim miasma over the area.[6]

Blending with it was the acrid odor of burned cabins, lean-tos, and corrals, residue of Kittredge's February attacks on the islands; an armed schooner he destroyed in July smoldered off Copano Bay. Gunpowder-sprayed estuaries, shrapnel-hammered sandspits, and crater-split mudflats seared Mustang Isle, as did chemicals the explosives released: vitriolic tartar, brimstone, and nitrate of potash. As acidic and caustic as the smoke from each blast, they parched the ground like salt over a field. And the bodies multiplied—mud creatures and baby fish belly-up in the bays, shorebirds scorched along the beaches, ground squirrels gutted by grapeshot. The lingering drought stripped the air of moisture and turned desiccated sedges and vines into tinderboxes.[7]

Meanwhile, the people of the Coastal Bend made ready. In early July, a squad reconnoitered Corpus Christi Pass, impounding boats and ordering islanders out. Additional soldiers entered the fort at Shell Bank, and a small garrison set up headquarters on the northern tip of Padre Island. Volunteers reopened the old Kinney fortification on the bluff and sank rubble into a trench partially connecting Corpus Christi and Aransas Bays. Residents accessed the city's arsenal—one eighteen pounder, two old smoothbores, and small arms accumulated over the past year—and checked guns. Finally, on July 20, they watched Major Alfred Hobby, a small cavalry detachment, and three hundred members of the Eighth Infantry Battalion join the meager group of horsemen and soldiers already in town.[8]

It was none too soon. By then, Kittredge had secured a yacht and a shallow-drafted steamer from New Orleans and a schooner and a sloop from the rebels themselves. On August 12, he steamed through Aransas Pass in the *Sachem*, unclogged the blocked trench in Corpus Christi Bay, and routed its protectors, taking possession of one gunboat and seeing two others set afire. Throughout the night, the rest of his makeshift fleet—the *Corypheus*, the *Belle Italia*, and the *Reindeer*—slipped over the bar, the confiscated gunboat *Breaker* in the rear.⁹ At nine o'clock the next morning, they loomed midbay as their lieutenant launched himself onto the long city wharf, flying a white flag and demanding carte blanche access to all government property.¹⁰

Kittredge's claim of access was as bogus to a city loyal to the Confederacy as was his Union rank, and—under the eyes of five enemy vessels and the citizenry of Corpus Christi—Major Hobby pointed this out. Even so, the lieutenant pushed on, demanding the withdrawal of Hobby's own men. He continued: there would be no damage to civilian property, but inhabitants had to "remove their women and children if they intended to make a stand." Once these words rang out, the waiting ended. Securing forty-eight hours to evacuate the town, Hobby turned to his troops, Kittredge sailed back to his steamer, and the populace panicked.¹¹

Women tore through kitchens, plunging kettles and utensils into makeshift bundles; children raced into alleys, chasing pets and livestock; daughters hassled youngsters into wagons, sons collected foodstuffs; fathers corralled mules. The city became a one-way exodus west, buckboards following buggies following carts following carriages.¹² And as their families moved into the plains, the men stayed on, anticipating action.

But there was none. A competent businessman, Hobby was an irresolute soldier, hidebound by precedent. Consequently, Kinney's old embankment on the bluff became battle central, its smooth-bores all the major had to repulse the twelve- and twenty-four-pound howitzers aboard the *Reindeer* and the *Belle Italia*.¹³ Beside themselves, veteran Billy Mann and engineer Felix von Blücher urged Hobby to reconsider. The solid-packed seashore, old home to Taylor's troops, they insisted, was a far more strategic artillery site. Honor bound, Hobby refused to do anything militarily during the truce, but once it ended, he caved. Allowed their head, Mann, von Blücher, and eleven volunteers laboriously pulled the old cannons off their plinths, edged them eastward, then eased them inch by inch down the bluff, pushing and shoving and hauling them toward the bay. By 2:00 a.m. Saturday morning the guns were sited; four hours later, they blew a hole in the mainsail of the *Corypheus*.¹⁴

Almost comically caught off guard, Kittredge and his sailors fired back, aiming desperately at the battery as blasts tore into *Sachem*'s side and rigging. Becalmed and too far out of range, the *Belle Italia* and *Reindeer*

Union ships in the bay, being shelled by Coastal Bend Confederates. *The Battle of Corpus Christi,* painted by Thomas Noakes. Courtesy of Murphy Givens.

lobbed shells onto shore, nine- and ten-inch cylinders spiraling uselessly through fences and deserted yards. They continued, however, sending over two hundred projectiles toward the rebels as Mann and his cohorts ducked, fired, ducked, and fired again.[15] Withdrawing deep into the bay at eventide, the schooner and sloop resumed their shelling at dawn as Kittredge frantically repaired rigging, patched *Sachem*'s hull, and seethed as another steamer went up in flames. Planning late into the night, he wrote to Secretary of the Navy Gideon Welles, "I hope you may be pleased with . . . tomorrow's work," then at dawn sent thirty men ashore to rush the rebels.[16]

Covered by a barrage of grapeshot and canister shells from the *Reindeer,* now finally near shore, Kittredge's men raced toward the battery, a twelve-pound rifled howitzer in their midst. Close enough to fire, their commander halted, positioned the gun, and then, accompanying the torrent of shrapnel and explosive balls now pouring from the *Sachem* and *Belle Italia,* "opened upon the battery . . . with precision" and deliberation.[17] But Hobby had not been idle. Closing upon the sailors' right, his small infantry attacked; then his cavalry, flailing cutlasses and bayonets, charged the melee. Outflanked and outnumbered despite the ongoing fusillade, Kittredge's men withdrew, leaving the seaside blistered with fire, iron balls, buckshot, and one defender dead.[18]

As well as innumerable chickens, domestic animals, and livestock as the lieutenant unleashed more rage on the town. "A great number of houses were shot through and through," a survivor wrote, "and a Newfoundland dog had its head torn off . . . bits of fence . . . were driven into house walls . . . all the hens were killed." Kittredge then pulled his vessels back, penned a self-congratulatory note to Secretary Welles, and "anchored in Aransas Bay . . . awaiting the return of the *Arthur* . . . before engaging in any more demonstrations."[19]

He needed no more. His boast, that he "held trade at his mercy," stayed true even when the lieutenant fell prey to his own enemies four weeks later. And despite holding him off, the people of Corpus Christi ended the year as defeated as if they had capitulated to Kittredge. "All the men from 18 to 50 years have now been drafted," one resident wrote in December. "The consequence . . . is that no field is tilled, and soon either flour nor corn will be for sale, and anyone who cannot live on meat must starve."[20] The Coastal Bend was becoming a wasteland.

But its barriers were still an avenue of access, and it was this that placed them at the crux of possibly the greatest invasion of the war, the Union advance on Padre, Mustang, St. Joseph, and Matagorda Islands.[21]

The intent was the capture of Confederate Fort Esperanza, which protected Matagorda Bay; the impetus was the series of mishaps befalling coastal Union forces after Corpus Christi's standoff. One, a December encounter on Padre Island, left officer Alfred Reynolds, leader of Kittredge's August landing force, severely injured. Another, a rebel attack on the ship *Sciota* on its way past Mustang Island, was almost farcical, *Sciota*'s commander simply noting, "I exchanged . . . shots . . . [then] continued my course." But the very effrontery of the exchange rankled commanders already smarting from losing Galveston in January.[22] It was General Nathaniel Banks's disastrous failure to take Sabine Pass at the juncture of Texas and Louisiana, however, that galvanized the North. By fall of 1863, with Vicksburg in Union hands and Gettysburg safe, Banks had a new plan: seven thousand soldiers and sailors were to convene at the most southern point of the Confederacy, take Port Isabel and Brownsville, and then conquer the coastal isles of Texas. In a final bit of irony, Banks put as commander the one man whose feelings about Texas—"This is the dirtiest place, I believe, I was ever in"—were never equivocal: Mexican War veteran Napoleon Jackson Tecumseh Dana.[23]

As knowledgeable of the territory south of Padre Island as he had been of Nueces and Corpus Christi Bays, Dana undoubtedly anticipated trouble once his men landed. What he may not have foreseen, however, was the late October squall line his steamer-borne troops encountered before they even got there. "On the second day [out of New Orleans]," one of his soldiers recounted, "the Gulf was quite rough, and the vessels of the fleet became considerably scattered. . . . On the thirtieth," he continued, "the storm became . . . tempestuous, the surf breaking over the vessels in a manner quite disagreeable . . . smaller vessels, especially, weathered the gale with extreme difficulty," as did the squadron's flagship. "The *Gen. Banks* . . . was but barely kept afloat. An ugly breach was sustained amidships, the cooking-galley smashed into smithereens, the steering-apparatus rendered unserviceable." Ignoring pleas to put in to rebel-held forts for relief, its commander did agree to lighten the hold. Tossed into the waves, then, were the very items vital to Dana's command: artillery equipment, battery wagons, harness mules. "Daybreak usher[ed] in a placidity of temperament as well as an unruffled sea," the chronicler marveled—such was Northerners' introduction to the coast.[24]

And such were their problems debarking: "The landing of the troops, horses, artillery, stores, etc., was accomplished only with extreme difficulty, after a couple of days of arduous activity," the writer resumed. "Many of the soldiers landed through a dangerous surf, boats frequently capsizing in the breakers on the shore, and a number of soldiers and sailors losing their lives." A far cry from the exuberant troopers jumping into waves years earlier, the experience may have reminded Dana, in the worst way, of General Taylor's barrier isle angst.[25]

But if the crossing was horrendous and the debarking treacherous, the commander had to have felt nothing but relief at his first encounter with South Texas rebels, for there were none. Convinced he could do more good defending the upper coastal islands, General Hamilton Bee had abandoned Brownsville and Port Isabel, leaving behind burning bales of cotton. There was nothing for it but for Unionists to commence their voyage up the Laguna Madre, vowing to learn "something of the country of which we had taken possession."[26]

And of the seacoast. Weeks earlier, Banks had sent a contingent of engineers in the steamship *Tennessee* to evaluate the barrier beaches and passes

Dana's men would be traversing. Having measured the depth between Brazos Santiago and Padre Islands, they moved northward toward Aransas Pass, remembering the earlier encounter with rebels the *Sciota*'s captain had reported. To their delight, the hidden battery at the tip of Mustang Island revealed itself in full armament when the *Tennessee* unfurled a British flag. "How are you rebs?" the Unionists chortled as the hoodwinked defenders scuttled out from their embankments like "bees from a hive." But it was their fortifications that got the Northerners' attention. "On the south shore of the pass," wrote a correspondent, "there are two batteries—one of four and the other of two guns. Five are twenty-four-pounders; the other is a rifled piece." He continued, "Three companies of infantry are encamped . . . behind a large sand hill. . . . Below their camp about half a mile are the hospital and powder magazine." Amazed at the sheer gaucheness of rebels who had considered island dunes sufficient cover, the *Tennessee* sailed onward. In its wake, chagrined Southerners wondered at the efficacy of their encampment.[27]

As well they might. The failure of the attempt a year earlier to set an artillery site close to Aransas Pass was a warning. Re-creating another at the most northern edge of Mustang seemed foolhardy at best, even using sand hills and coppices as cover. Moreover, there was insufficient timber available upon which to base the cannon. Yet Fort Semmes had become reality, its name honoring the South's most notorious sea captain, and its men pledged to protect the pass that had become known as blockade runners' "mine of wealth."[28]

But even their determination was suspect. Rumors of Banks's invasion plan had already seeped into Southern communication channels by July, and it was during that month that troops originally stationed at Fort Semmes were transferred out, to be replaced by older conscripts and horsemen. Their battalion colonel was so incensed at the assignment, he resigned, protesting "the stationing of cavalry upon those islands." By November of that year, the contingent had decreased yet again, numbering no more than forty-two officers and men within the encampment.[29]

But even more unsettling had to have been the conflicting demands the commander of Fort Semmes began to receive. In August 1863, just after the new draftees and horsemen had been garrisoned, Major General

John Magruder in Houston ordered "the fort at Aransas . . . discontinued" and that "'Quaker guns' [fake cannon] be placed in the most conspicuous position."³⁰ Within a few weeks, however, he reversed himself, sending state troops in to replace volunteers in the now-not-to-be-deserted fort. By October he had reversed himself again, ordering Fort Semmes to be evacuated and its operating guns sent north. An attempt to obey this final command was stymied by high tides along the coast, and by November Fort Semmes remained occupied, armed, and effectively ignored by mainland rebels.³¹

Yet it was open to the enemy, an incessant, oncoming army with little regard for the layered soils it was upheaving. Dana's troops had been spared a trek up Padre, "about sixty miles long . . . elongated . . . and very unattractive," in favor of steamer transport to the southern shore of Mustang Island. From there they had offloaded guns, carriages, mules, and equipment onshore, re-formed their positions, and then started walking, trudging nearly twenty miles forward over hard-packed sand. Finally, just short of "the fortified position held by the rebels upon the northern extremity of the island," they stopped, grimy and sweaty even in the cool November morn. As they paused, however, Union skirmishers crept ahead, and almost at the same time the troops resumed their march, shots came from Fort Semmes pickets. "Our advance at once opened fire," a soldier remembered, "the gunboat *Monongahela* . . . tossed a few shells into the enemy's works," and the armed transport *McClellan* blasted more. Forming a storming party, Union regiments "were moving steadily forward . . . when they confronted a party of rebels bearing a flag of truce . . . [with] an offer of surrender, without conditions."³²

The defenders of Aransas Pass had capitulated, "nearly one hundred stalwart Texans," a Maine veteran remembered, "tall, stout, robust-looking fellows . . . with heavy guns, a large quantity of stores, and munitions of war." Later chroniclers recorded three cannons, "serviced by . . . nine officers and eighty-nine enlisted." Accompanying the seizures was a chart, confiscated from the rebels and subsequently inscribed, "Aransas Pass, Texas: Taken by the Union Forces under Major General Banks, November 17, 1863." Included on it were instructions on crossing the bar: "Bring the Rebel Works on Mustang Island to bear west by south, and to within 50

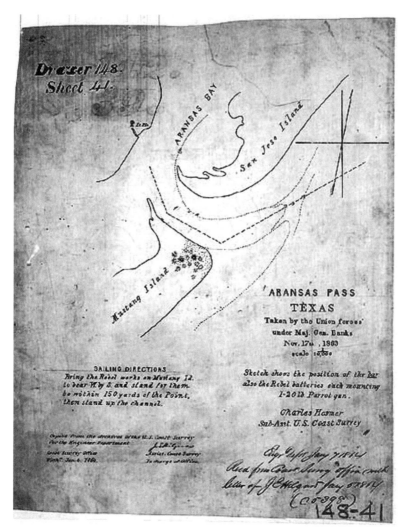

Fort Semmes battlements at the northern tip of Mustang Island. *Aransas Pass Texas, Taken by the Union Forces under Major General Banks, Nov. 17, 1863.* Courtesy of the Port Aransas Museum.

yards of the Point, then stand into the channel." But perhaps the most telling aspect of the surrender of Fort Semmes was a Union general's later assessment. "The works on Mustang Island are of little or no value . . . from [their] impracticality . . . amid . . . shifting sands."[33]

The barrier's shifting sands were not the only reason for the fort's impracticality; its cannon pointed only toward the pass. The very concept of an enemy approach from behind had never been considered. Such good fortune—or misjudgment, depending on one's point of view—held true for Dana's men as they "tarried at [the fort] for nearly a week," then crossed the pass to St. Joseph Island. Their luck held there as well, the isle being "a veritable oasis in a great sandy desert," one trooper remarked. "Here was to be found excellent grazing land, with attendant fine herds of cattle . . . sleek and plump, and ever ready at hand." As profligate with their ammunition on livestock as Taylor's troops had been on wildlife, Unionists left grasslands and ponds peppered with minié balls as they marched northward.[34]

But indiscriminate destruction of island life, managed or natural, was not to go unavenged. The troops felt nature's first retaliatory blow shortly after a fatal encounter with rebels.[35] Approaching the pass to Matagorda Island—"deep and something like three hundred yards wide, with a very strong current"—they encountered a norther. "A fierce gale [came] sweeping down the Gulf—angry waves . . . dashed upon the beach with great velocity . . . loose sand . . . scattered in vast clouds, in every direction, toppling over tents, filling eyes, ears and nostrils. . . . The only resort of our soldiers was to burrow in the sand, in pits . . . and [lie] covered with raw hide from cattle recently butchered."[36]

Nor was the experience a one-time occurrence. Finally crossing Cedar Bayou by stringing shallow-bottomed boats together, the Union troops marched until they got within sight of Fort Esperanza—"[walls] anywhere from ten to fifteen feet high, and at least fifteen feet thick"—when "another of those unearthly northers sprang up." Despite the raging blasts, fatigue parties from each regiment began digging battery pits parallel to the fort, and by dawn, "our artillerists . . . commenced to drop shells into the enemy's stronghold." An incessant barrage all day compelled the rebels "to seek refuge within the inner works of the fortress," and that midnight, "a terrific explosion was heard within the fort." The rebels had blown up their powder works and, using the blinding gales to protect themselves from naval vessels offshore, had fled into the mainland of Texas.[37]

With their abandonment, Matagorda Island became Northern territory, as had Padre, Mustang, and St. Joseph Islands. The Unionists did not stay long. By June of the next year, most occupation troops had withdrawn. Some few returned at war's end, however, "to demonstrate their presence to the Confederates . . . at strategic locations along the Gulf

Coast." Like their Mexican War cohorts and more recent veterans, they soon felt the wrath of the isles: "Marshall Montgomery's company was dispatched to Mustang Island for a . . . few months," a descendant later wrote, "living in tents on the beach, coping with sharks, mosquitoes, and big red ants. . . . As the ants came in procession from the beach, he had his bed covered with screen cloth and the bed posts set in cans of water. One night, they discovered a hole . . . and he was forced to run for the ocean, clothing in hand."[38] The military's encroachment on the barrier isles was not to go unchallenged.

PART III. ENTREPRENEURIAL DAYS

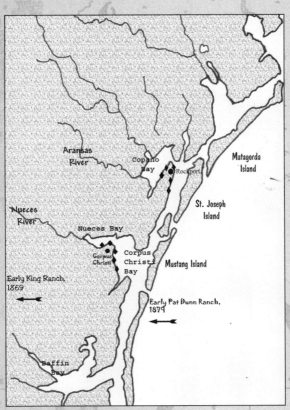

Meat-packing houses and early King and Dunn Ranches. Based roughly on Murphy Givens and Jim Moloney, *Corpus Christi: A History*; Tom Lea, *The King Ranch*; and Greg Smith, interview and presentations.

CHAPTER 7

Herders and Hiders

The military would return, however—the army as well as the navy—to buttress and bind those barrier isles. But that lay in the future. It was a bedraggled present that confronted ex-Confederates by the late 1860s—a present personified by the multicolored, long-legged, tick-carrying cattle that surrounded them.

Descendants of Iberian herds brought centuries earlier by Spanish colonists and missionaries, these longhorns had thrived in the brush country west of the bays. There, their elongated snouts chomped on low grasses, and their tough skin repelled spiny shrubs and cacti. Inbred ferocity and rapier-like horns repulsed predators, and a seemingly incessant sun promoted endurance. Females calved disease-resistant offspring on a regular basis, while males scouted treeless plains for sustenance. Lean, muscular,

Texas longhorns near Corpus Christi. Photographed in the early twentieth century, these rangy cattle retained several of their ancestors' most belligerent characteristics. *Above*, courtesy of Corpus Christi Public Libraries; *right*, courtesy of Jim Moloney.

and feral, these bovines bore little relation to the docile Herefords that had populated the eastern Atlantic.¹

But those Herefords were gone, slaughtered to feed Union armies during the war. Still present, however, was easterners' need for leather and beef, a need longhorns could now assuage. And so began in earnest huge movements of Texas cattle, heading north from the Nueces River. Ramrodded by ex-rebels and staffed by waddies, vaqueros, and blacks, the trail drives were a measured gamble by ambitious landsmen to make money from their meager surroundings.² And through their efforts, some found themselves in thrall to the islands.

Richard King was one. Eastern born of Irish parents, he apprenticed early to a New York jeweler. But the sea called, and before King turned twelve, he had stowed away on a schooner bound for Mobile. Captured but accepted into service, the youngster soon gravitated to riverboats and, by the age of sixteen, had so mastered the skills necessary to maneuver paddle wheels that he earned a pilot's license. Time spent with the military ferrying troops during the Seminole Wars brought him into contact with another riverman, Mifflin Kenedy. By 1847, the two had joined Zachary Taylor's army.³

A year after leaving Corpus Christi Bay and blooded with victories at Monterrey and Buena Vista, many of the general's troops were now south, with Winfield Scott. Yet Taylor still had to maintain a defense, and it was his depots along the Rio Grande—Reynosa, Camargo, and Mier—that King and Kenedy were ordered to supply. Loaded carefully from ships at Brazos Santiago, their boats were to steam down to Baghdad, slip across the bar at Boca del Rio, and then chug carefully upstream, edging around the tributary's fallen trees, abrupt bends, and sudden eddies. Erratic at best, the Big River had flooded months earlier and by that summer was so mud filled and shoal infested that even seasoned pilots were running their vessels aground.⁴

But King and Kenedy managed and, within two years of war's end, had such a thorough knowledge of the Rio Grande that they started their own shipping company, a company that, in time, virtually monopolized river freighting. The incessant demand for repairs, however—hulls that needed caulking, propellers that demanded realignment, boilers that threatened explosion—preyed on the young captain. Just as frustrating were barrier isle beachings where King and his crews worked for days wresting stalled steamers off sandbanks. "Boats—they have a way of wrecking, decaying, falling apart, decreasing in value and increasing in cost of operation," he observed years later. "But . . . cattle and horses, sheep and goats, will reproduce themselves into value."[5]

Consequently, King slowly divested himself of steamboats. Within four years of Lee's surrender at Appomattox, he could claim title to over sixty-eight thousand acres of South Texas brushland, at least twenty-three thousand head of cattle, a series of hide-and-tallow factories, and one piece of Padre Island.[6]

The sixty-eight thousand acres of brushland made sense. As semiarid and sparse as the area was, Captain King (as he continued to be known) needed a lot if he wished to graze herbivores like cattle and horses—and he knew it. "Buy land and never sell" became his watchword, and his holdings grew, as did his herds.[7] He took one thousand of the approximately twenty-three thousand cattle he owned in 1869 to St. Louis on the first King Ranch trail drive that year. He put thousands on the road in subsequent years and made thousands of dollars more, even allowing for expenses and shares to his partners. One expedition alone, completed and tallied in November 1875, netted the captain $50,000.[8]

And from his eminently practical point of view, King's investment in hide-and-tallow factories made sense as well. Just as boats decreased in value over time, livestock could lose their buyers—and had in the Panic of 1873. Moreover, cattle skinners had viciously demonstrated the efficacy of simply slaying bovines on-site, stripping them, and selling their hides to the nearest purchaser.[9] In an age of fluctuating markets and disreputable distributors, why not slaughter one's own cattle and forgo the eastern middlemen?

But where to put the holding pens and killing stations and stripping areas for such an enterprise? Where to set the fat-rending vats? Where to locate the curing racks? The answer, for Richard King and others, lay in the "sweet . . . bright waters" of the bays, so extolled by Taylor's men thirty years earlier. For upon Corpus Christi and Aransas Bays lay wharves, and from wharves, a new kind of cattle drive could sail forth.

And it did, as packing houses of every size and description began to encrust the shorelines. Corpus Christi had endured "gangrene[d] hides" left in open air earlier, and in the years before the Civil War, a small packery had aired its stink on the peninsula north of the city. By the 1870s, another had been set up on that North Beach peninsula, one was operating close to Oso Creek, one was on Mustang Island, and King's was right on the bay, just south of Corpus Christi proper. There the captain kept it until town fathers made him remove it for health reasons. But he maintained others, visible testaments to the rancher's hard-nosed practicality.[10]

Why, then, did the man purchase an island, or at least thirteen thousand acres of it? Why extend his brush-filled pasturage onto mudflats and sand dunes? Why own a section of solid silt where the only fresh water hid underground? Padre Island did host vegetation; pepperwort and seedbox grasses, beautyberry and yaupon bushes, water lilies and cattails grew in its swales. But longhorns were not native to this isle; to be profitable, they would have to be waded onto it, then off again.[11]

The captain never really explained his action and later appeared to regret it for the complications it entailed.[12] Some assumed he made the buy as a good deal, costing a little over one cent an acre. Others thought King had planned to establish a major port in Corpus Christi, for which Padre would be a suitable holding point. A few speculated that he had hoped to ship southward, to Cuba.[13] But none seemed to wonder whether his purchase of land on this barrier island was the stowaway boy's last chance to glimpse the sea. It surged there, on the eastern side of the isle, its waves crashing the beach and its winds shifting the sand. Only a few miles away from the desert shrubs of South Texas, Padre Island may have brought the ocean back to Richard King.

It brought survival to Patrick Dunn, the second landsman entranced by the isles. Just a child when Kittredge's ships shelled his town, the youngster watched his father die not long afterward, then witnessed his mother's ongoing struggle to keep the family alive. It was a move to Padre Island in 1879, however, that galvanized the youth. Separated from restraints on land, he studied the isle and made it work for him. Realizing that fresh water pooled above salt in groundwater deposits, he sank "Ol' South Tanks" (troughs with wooden sides and no bottom) at the base of dunes and left cattle to drink. Aware that incoming tides provided weeds and sea grasses, he free-ranged his herds and allowed them to forage the shoreline. Surrounded by varying widths and depths of the moatlike lagoon, he restricted access to visitors and ordered interlopers off.[14]

He kept cattle on by herding them carefully across the shallows between Padre and Encinal Peninsula. Dunn then allowed his animals free range, using a fence only in the southern area, where grass dwindled into hard beach and shells. He did not coddle his livestock but did eschew roping. His calves were "mugged" instead, cornered in the small chutes he had built of driftwood, and thrown sideways to the ground to be branded.[15]

Roundups took three to four weeks and started at the southern tip of the island. From there, vaqueros and waddies herded the cattle northward, stopping every evening at one of the strategically located line camps Dunn had erected. Then livestock were penned and men were fed and rested in preparation for the next day's drive. At its end, just south of the farthest reach of the isle, the rancher and his men herded the cattle westward to the mainland, recrossing through the Laguna Madre shallows.[16]

As easygoing with his herds as Dunn had been on the island, however, old conflicts resurfaced once he was ashore. One involved the mainland. Although the earlier enthusiasm for hide-and-tallow plants along Corpus Christi Bay had lessened, the craze had spread north. A protrusion on Live Oak Peninsula, bordering Aransas Bay and notable for its barrenness, had now become headquarters for one major packer and was accommodating more. William Hall built a factory there, as did William Wood and the Coleman-Mathis-Fulton Company.[17] Before long, the new enclave, Rockport, was sporting as many cattle pens, slaughterhouses, curing sheds, and drying frames as had its older sister downstream—with as much decay. "The slaughter of animals was so great," one Corpus Christi chronicler wrote, "that the packers did not take time to cure the meat, and the odor . . . became almost unendurable."[18]

The cattle themselves died as they trod down narrow chutes attached to holding pens, speared through their necks by trained "stickers." Immediately dehorned, decapitated, and deskinned, they were piled—heavy and bleeding—upon platforms until finally quartered and heaved into tanks. Boiled until their fat, whitish and coagulating on the surface, was skimmed off and their bones lay broken on vat bottoms, the remains were eventually dumped into the waterways of the area. Such carrion became seductive nutrients to certain marine life forms of the Coastal Bend—and a cause of their own near obliteration.[19]

But for the landsmen of South Texas, the hide-rendering process was uniquely appropriate for the area, one that no entrepreneur could ignore. Dunn certainly did not, nor did he evade his options: drive the herds to such packeries for slaughter, head them to rail centers north and later west, or turn them toward the docks at Rockport for shipping upcoast.[20] Every

Marion Packing Company, located north of Rockport, in the 1870s. Painting by John Grant Tobias. Courtesy of Texas Maritime Museum.

choice meant profit. The nearby plants seemed to promise more, however, since they were closer and their products were gratifyingly tangible.

Tangible and prompt and popular: hide-and-tallow products burgeoned in that era of seaside slaughter. In 1869 over a million pounds of dried hides and close to the same number of wet salted hides were generated in Aransas Bay, as well as nearly 400,000 pounds of tallow and 90,000 pounds of bones. Three years later the warehouse master in Corpus Christi listed 130,000 dry hides ready for export—along with 19,000 others still damp, 2,000 barrels of tallow, 29,000 horns, and over 140 tons of bones.[21] But being clustered in depositories did not make these items profitable. Getting them to market did, and only one man could do that for men like King and Dunn: Charles Morgan.

He was the embodiment of South Texas shipping by the end of the Civil War. One of the first to see potential in the western part of the Gulf, Morgan delivered troops and mail for Taylor during the Mexican War, and passengers and freight thereafter. His New Orleans–Galveston route became classic, and eventually Indianola, St. Mary's, and Brazos Santiago became part of his runs. He even made connections with Richard King's riverboat line up the Rio Grande. But it was the Coastal Bend's hide-and-tallow factories that brought Morgan's steamships to the brand-new settlement on Aransas Bay in 1867. By the end of that year, Rockport was as important to the Morgan Shipping Lines as New Orleans.[22] But neither was as valuable as the third landsman enrapt by the isles, Robert Ainsworth Mercer.

CHAPTER 8

Helmsmen

Mercer's decision to become part of South Texas had not been taken lightly. Born and bred in a river-bordered region of northwest England,[1] he had taken his wife and two surviving sons to another in Indiana in 1830. There, near a major bend of the Ohio, he made his living in Floyd County, close to the largest shipbuilding center in the state and, by the early 1800s, home to steamboat production. For the next twenty years Robert and his wife, Agnes, raised children and crops.[2] By the time they moved once again, to Mobile, Alabama, his eldest son had become a ship captain, his third born owned his own boat, and his fourth surviving boy worked on the steamer *Cuba*.[3] It was his move to the Coastal Bend in late 1852, however, that brought Robert Mercer home. Within three years, he, Agnes, and their remaining brood had built a wharf and a warehouse on the northwestern tip of Mustang Island and, right next door, a new farmhouse lovingly christened El Mar Rancho.[4]

"We are all well," Agnes wrote her oldest offspring in May 1856. "We have . . . provisions to last until Christmas . . . plenty of milk and butter . . . and beef, pork, and mutton for life. . . . We have just done some sheep shearing," she added, and "had a merry time of it." In a tender afterthought, she finished, "Father and all send their love."[5]

The cheeriness would not last. Kittredge's destruction of their home, livestock, and corrals drove the family off Mustang six years later. When they returned at war's end, Agnes was no longer alive. Nor were Robert's two sons who had stayed in Mobile. Peter, the eldest, died after being captured by the North trying to slip through the Atlantic blockade, and William Henry disappeared as did his steamer *Cuba*, burned by fellow rebels to prevent a Union seizure in 1863.[6]

But the family endured, and it was Robert's work on the island of Mustang and within its waters that not only kept his people alive but bolstered the entire economy of South Texas. For they had joined that exclusive

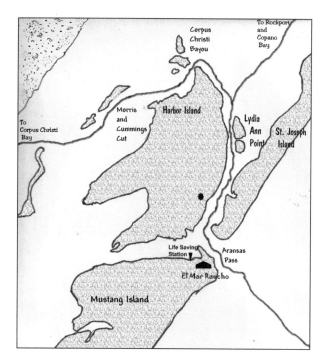

Generally used routes to Rockport or Corpus Christi. Based roughly on J. G. Ford, *The Mercer Logs*.

band of seamen extolled by the federal government as "unique in the maritime world": bar pilots.[7]

A later Supreme Court decision put into words the common consensus regarding pilots: "[They] have an independence wholly incompatible with . . . general obligations. . . . No person can tell them how to perform their pilotage duties. . . . [They] are free to act on their own best judgment."[8] It was the same kind of judgment that Diego Miruelo had used so poorly in 1528 when he offloaded Pánfilo de Narváez and his crews onto the peninsula of Florida; it was the same kind of judgment Juan Almonte admired three hundred years later when he encountered Coastal Bend Karankawas for the first time. Used wisely or not, a pilot's judgment was an accumulation of experience, observation, and environmental accord. This was the kind of judgment Mercer attained when he settled on Mustang Island, and it was the kind he maintained when he returned; he and the isle had bonded.

The island's waves were hypnotic, soaring higher and higher across the rising seafloor, then crashing downward; its currents swirled in season. Its storms thrust silt westward and its sediment occluded coves. Natural accretion shifted its buildings inward and extended its reefs.[9] Its ducks outwitted young hunters, flying off "before the boat got a half mile of them." Its beaches exhausted foragers and its insects confounded homegrown exterminators. Its wild mustangs stampeded past captors "as if hell kicked them on end," and its rain showers "made the grass look green" again. Even the occasional blizzard—"it piles the snow up in ridges in the lea of the house"—was delightful. His son reported, "Father . . . takes a short cut . . . to feed . . . the calves: goes through the window."[10]

But most of all, Robert Mercer gloried in his island's bar-crossing chores: setting out tall range poles to direct incoming vessels, racing to Mustang's edge to guide in a schooner, entertaining a captain while awaiting a lifting tide. The chance to take responsibility for a ship and bring it through the pass delighted him, as it did his sons. By 1866 the former Englishman was

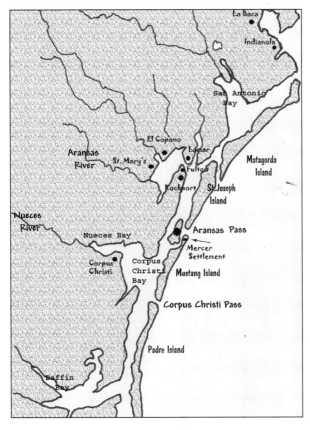

Port towns and their passes in the mid- to late 1800s.

License granted John Mercer in 1891, appointing him branch pilot of Aransas Pass. Courtesy of Port Aransas Museum.

appointed wreck master of the pass by Governor James Webb Throckmorton, and by the 1870s sons Ned and John were bar pilots too. They joined a rough-and-ready group, independent of bosses, "irregular . . . rugged . . . and hard-drinking," but committed to protecting the pass and those who traversed it.[11]

Traversing the pass meant revenue. For Charles Morgan, having an experienced guide convey his schooners over the bar meant they could sail inland enough to pick up loads near St. Mary's or Fulton. For the shipping magnate's steamship captains, it meant they could access even heavier cargo by having pilots maneuver them through the pass, head up along Harbor Isle, and venture their boats just far enough into the bays to dock at Rockport or Corpus Christi. At either place, nail kegs filled with silver dollars were exchanged for drums of whitened bones, barrels of tallow, bundles of hides, and batches of horns, hooves, and teeth. The trade was so brisk it even garnered the governor's attention: by 1874, 32,000 skins, 2,000 hogsheads of lard, 8,500 barrels of salted beef, and 64,000 sets of horns had cleared Rockport—carried on vessels guided by the Mercers and their fellow pilots.[12]

At least most were. Not all captains chose to hire professional expertise. "Schooner *Anna Hanson* got underway without a pilot," one of the Mercers wrote in 1871. "She did not go far before she fetched up on Lydia Ann Point. So much for next time she [goes] on in such a hurry." Other captains were simply stubborn. "Ned and Clubb went up to the [schooner] *Whisper*. The Captain would not go out." When that happened, there was just one remedy: "Came home and planted potatoes."[13]

For piloting was unpredictable. January storms could stymie missions. "Steamer off the Bar," one Mercer wrote in 1873, "but the weather was too thick, lost sight of her." Or competition could cut one out. "A vessel hove in sight from southward. We started. Captain Clubb started. Had a race but Clubb was too much for us."[14] Rushing the pass to secure business was wise; the standard fee for guiding a vessel was four dollars for each draft foot.[15] When even Morgan's lighter steamships drew seven feet of water, a commission of twenty-eight dollars was a good day's work. In time, vessels refusing a pilot's aid still owed half fees, and there were other ways of profiting from the big ships.[16] Lighters—those shallow-hulled boats that had carried so much contraband within the bays during the Civil War—were still necessary for transferring cargos from steamers too impossibly huge to clear the pass; their crew and operators made money too.[17]

Profit was important. The state fielded some responsibilities for navigation in the pass, and personal income helped defray expenses—the Mercers raised crops and livestock to supplement their guiding fees.[18] But maintaining safe passage over the bar demanded specialized equipment: small boats to carry the pilots to approaching vessels, a solidly constructed wharf with davits upon which to suspend those boats for repairs, at least one supply raft to offload personally purchased cargo, a warehouse big enough for salvage storage, a pilot house specifically designed for land-and-sea communication, and various sailboats or scows for intralagoon transportation.[19]

Moreover, items vital for navigation were costly. Some of the earliest devices were signal poles or range markers, tall stake-like structures placed in water and on land. Their location helped indicate channel boundaries and enabled ships to "sight" their bearings: when the vessel was where the pilot could line two markers up visually, it was practically guaranteed a safe course. Correct siting of the flagpoles was vital—a careless or hasty displacement could ground a boat. It definitely brought curses upon the miscreants. "Heath . . . sneaked down to Mustang and put up his ranges . . . without sounding the bar. . . . He is a low down sneak," one of the Mercers commented in 1874. One month later, the incident was repeated. "Schooner *Alice Taylor* arrived from Calcasieu. Set signals for her to come in. Old Roberts, the damned sneak, watched where the signals were put, then he put ranges for the steamers to go out. He did not know how much water [was in the pass]. No more than a hog. He is a damned hound— worse than Heath."[20]

Equally as vital to correct range marking were lead lines, little different from those used to measure depths by Diego Miruelo in the 1500s. Still as basic as a metal alloy tied onto knotted cord, dropped overboard, then hauled up and "read"—the number of wetted knots indicating the number of feet to the bottom—the lead line "soundings" were indispensable to pilots guiding vessels and far cheaper than the mechanized devices distributed by most navies.[21]

As necessary as range markers and lead lines and used almost as long in Gulf waters were buoys, "stationery floating signs . . . anchored to the bottom of the channel by cables." Their shapes, painted signs, and illuminated lenses signaled mariners by day or night. Some had bobbed in the currents surrounding Aransas Pass since Taylor took his troops to Mexico, but by the 1880s, more were needed. Within two years of an impassioned plea by the *Corpus Christi Caller*, the outer bay buoy was secured a mile and a half off Mustang Island and a red "nun" buoy hovered nearer shore, its oval shape reminiscent of the garb Roman Catholic sisters wore at that time.[22]

But despite the practicality and availability of such devices, nothing eclipsed the importance and expense of flag-and-light indicators. A pilot's expertise in their use—as inscribed by the British in their 1874 Code List of Signals—was an international requirement. Adopting these signs for their merchant marine as well as their naval forces, maritime countries had mandated they be "the universal means of communication between the Ships and Signal Stations" of all nations. As a result, Aransas Pass pilots joined those around the world in recognizing distress signals used by ships: during the day, a gun fired at intervals with the International Signal of Distress flag (often a square flag with a ball above it) raised, and at night, guns fired or rockets shot periodically.[23]

More usual displays were those hoisted by vessels on their front mast or foredecks when they needed pilot expertise at the pass: the national flag (for Morgan's ships and others, this was the Stars and Stripes surrounded by a white border) or the International Code Pilotage Signal flag (if not the national flag, this was a ball between two pennants). At night, a blue light flashing every fifteen minutes or a bright white light located just above the bulwarks, blinking every other minute, let Mercer and his sons know a ship needed passage.[24]

Although such lights and flags were standard for ships, the message they imparted was as necessary to an Aransas pilot as knowledge of the pass itself. Consequently, the Mustang men studied the flags' sizes—six to eight feet wide. They memorized their shapes—swallow-tailed burgees, solid-colored

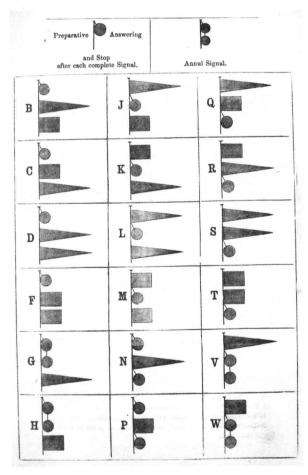

Letters of the alphabet, as communicated by flags, pennants, and streamers. John J. Mayo, *The British Code for 1874 for the Use of Ships at Sea, and for Signal Stations.* Courtesy of Port Aransas Museum.

pennants, and striped square streamers. They learned their meanings—a single pennant C displayed meant "Yes," a single pennant D displayed meant "No," and a swallow-tailed flag flying above three standards indicated geographical directions were coming. Long use and practice with signal flags could convey a whole conversation between ocean and shore—a boon to those landbound but desperate for news of loved ones at sea.[25]

Such was the duty of pass pilots—to learn the international language of mariners. It cost. Tests, certification, and licensing levies mounted up, as did the purchase of works like Mayo's *British Code List*, the equivalent of one-fifth the fee for bringing a light steamer across the pass.[26] Just as expensive were the flags local guides positioned at the pass to direct client ships. If shoals and sandbars were too obstructive, the signal flag was white: "Bar not passable." If the tide needed to come in, it was blue: "Wait for high water." Favorable conditions for crossing the pass warranted a blue range-marker flag in the front coupled with a red flag in back.[27]

Bar and Bay Pilots Association members even used flags to establish an incoming ship's draft—and thus determine their fee. Once the vessel had raised its required signal flag ("Want a pilot. Can I have one?") and been answered by a C pennant on the island signal staff ("Yes"), it set its signal halfway up the mast nearest its bow to indicate six feet, six inches of draft. If the boat needed seven feet, three inches of water for clear passage, its flag flew midway up the central part of the foremast. If its draft was seven feet, nine inches, the flag was at half-mast on the ship's principal spar. For even deeper draft, the flag topped the main peak, which it did on a regular basis when the steamer *Aransas* came through the pass.[28] For those boats unable to ever clear the pass, the association even provided its own lighter.[29]

Expenditures were significant, therefore, for those who crossed between Mustang and St. Joseph Island and for those who helped them do it. But just as important was safety. By the late 1880s so many vessels crowded harbors like Galveston's or Lavaca's that ships began sporting direction indicators to prevent collisions. Such traffic was understandable. Matagorda Bay's most prominent port, Indianola, had ceased to exist following two decades of hurricane damage, so prudent townspeople were moving farther in. "An effort is being made to change the terminus of the Gulf, West Texas, and Pacific railroad from Indianola to Lavaca," one local paper reported after the most recent hit, "[and] . . . the county seat of Calhoun

CODE OF SIGNALS,

TO BE USED

While Crossing Aransas Bar.

FOR STEAMERS AND SAILING VESSELS.

WHITE FLAG, Bar not passable.
BLUE FLAG, Wait for high water.
These signals will be hoisted on the Signal Staff on Mustang or St. Joseph Island.

RANGE FLAG.

Front Range, **BLUE FLAG.**——At night, **WHITE LIGHT.**
Back Range, **RED FLAG.**——At night, **RED LIGHT.**

When there is water sufficient to cross the Bar, the range flags or lights will be set as as above. The ranges will be either on St. Joseph or Mustang Islands.

Vessels running for the Bar with signals set for a pilot, will be answered by a signal set on the signal staff. The vessel should then haul its signal down, and set it for

THE NUMBER OF FEET THE VESSEL DRAWS,

BY THE FOLLOWING SIGNALS:

For 6 feet, 6 inches, Flag at half mast on Foremast.
" 7 feet, Flag at Foremasthead.
" 7 feet, 3 inches, Flag at half-mast on Foretopmast.
" 7 feet, 6 inches, Flag at Foretopmasthead.
" 7 feet, 9 inches, Flag at half-mast on Mainmast.
" 8 feet, Flag at Mainmasthead.
" 8 feet, 3 inches, Flag at half-mast on Maintopmast.
" 8 feet, 6 inches, Flag at Maintopmasthead.
" 8 feet, 9 inches, Flag at half way up to Mainpeak.
" 9 feet, Flag at Mainpeak.

A GOOD LIGHTER, of five hundred barrels' capacity, will be held in readiness for lightering vessels outside or inside the Bar.

C. C. HEATH,
WM. R. ROBERTS,
Bar and Bay Pilots.

ARANSAS PASS, August 1st, 1874.

C. A. Beman, Prin r, Valley Office, Corpus Christi.

"Code of Signals" provided by the Bar and Bay Pilots Association for vessels approaching Aransas Pass. Courtesy of Port Aransas Museum.

County is also to be moved back to Lavaca." Consequently, the tiny city of Lavaca, long eclipsed by its eastern rival, had become a dominant port in the central Gulf Coast, basking in its double-pass access, Cedar Bayou to the south and Pass Cavallo to its north.[30] Even prouder—although less protected—was its congested counterpart farther north, Galveston. Occupying both sides of an island, with a natural harbor to its west and open coastline on the east, Texas's "Queen City" hosted one of the foremost cotton ports in the world and considered itself "the most advanced and sophisticated" city in Texas. Yet its prestige came at a price; the barrier's location was so vulnerable to storm surges and current shifts that whole islands in its past had disappeared—Malhado, upon which Cabeza de Vaca and his compadres had landed, was now a peninsula. Similarly, the sediment-formed ridge underlying the passage between Bolivar Peninsula and Galveston Isle was beginning to broaden and expand. By the 1870s, it had formed two bars, each of enough density to stymie the shipping so vital to the bay. Determined to avoid blockage, business interests ushered in the federal government, begging the same engineering expertise that bombarded the city during the Civil War to bolster it now.[31]

No such interest drifted south, not yet, and solitary Aransas stood unfettered, its shoals still shifting, its breakers off St. Joseph still catching captains off guard, and its submerged sandbars still twisting propellers awry.[32] The pass challenged human-made vessels at every crossing attempt—and quite often won.

As it had in December 1866, grounding the schooner *Adair* even as it approached. "She was gone up very heavy," Mercer wrote later. "Got the water out of her but could not [get her off] the riffle." He continued, "Worked at her all day but could not do anything for her. Moved her a couple of lengths but she leakes [sic] so bad that she cannot be set free. We had to give her up for a bad job."[33]

"A stiff gale" did similar damage to the schooner *Surprise* five years later, this time tearing it apart before washing its remnants "among the sand hills." But it took the isles' oldest weapon, a November norther, to destroy one of Morgan's proudest steamships, the SS *Mary*. "It struck the vessel," a passenger remembered, "the night we reached the bar," and by morning, "water . . . was pouring through the ship like a mighty river." His vessel obviously damaged, the captain tried to enter the pass anyway. "Struck and failed," a Mercer reported laconically; "she is a total loss."[34] But here the inbred dedication to safety, part of family tradition, came to the fore. "John, Ned, Tom, Tomie and Perry went . . . alongside. Took the passengers on shore and all hands from the ship . . . her fires were put out in thirty minutes afterwards. . . . No lives lost." Cleaning up the damage continued for weeks: "The *Mary* is breaking up. All of her upper works is coming on shore. The beach is strewed with logs. . . . All hands cruising the beach . . . our squad hard at work. . . . So ends the day."[35]

Removing wreckage, clearing the pass, securing damaged cargo for insurance inspectors—all were part of the wreck master's job allocated to Robert in 1866. Just as vital was his son John's appointment later on, when he became head of the first lifesaving station on the isle.[36] Along with maintaining up-to-date equipment and training his crew, he also had to know the weather, and by 1880, weather information, utilizing the medium of telegraphy, was on the rise. Ever since the government had established its first observatory in 1871, the practice of using daily temperature readings, barometric pressure measurements, rain gauge amounts, and wind direction data to make reasonably accurate forecasts spread. By 1873, farmers all over the nation were being informed as to what was coming, as were sailors, shippers, and pilots—all part of the duties of the federal government's Signal Service Corps.[37]

But the weather had always been a natural part of Mercer life. In his January log of 1867, Robert recorded "a snow storm on the night of the 1st and early all the 2nd. It was awful . . . coldest time ever seen in Texas . . . [could] hear my horn toot." Eight years later another January brought "a regular peeler, with plenty of ice. . . . The bay shore and all the ponds are froze over. John and Frank walked across the Cove; there was ice all the way across." Other years brought no easing: "Oh Jehosophat but it is cold. . . . a regular peeler with ice all round . . . and plenty of hailstones to make it interesting."[38] Extreme temperatures were as innate to the island as abnormal tides. "5:00 PM, the wharf underwater, except for the T-head," wrote one Mercer in September, adding a few days later, "The tide extraordinarily high. Moved the boats farther back to keep them from washing away. . . . The wharf from the T-head . . . all washed away and came ashore. . . . Tide still booming high." Such overwhelming inflows as had kept soldiers from evacuating Fort Semmes in 1863 remained an isle attribute, as did wind. "This day begins with the wind West," Mercer wrote one February, "a regular peeler. All the seeds that John and Tom planted are up and gone towards the beach." The next year it came from the "north-north-west, blowing as hard as it can . . . the front bar range washed down and came in the Cove."[39]

Not all wind damaged. One in April 1876 was "fresh . . . a good breeze," and that September the Mercers were able to celebrate the end of the

Ruins of Indianola after the 1886 hurricane. Courtesy of Victoria Regional History Center, Victoria College / University of Houston–Victoria Library.

month with "no storms, nor the wharf wash[ing] away. We are having something singular," the writer remarked dryly, "for the wharf got in a habit of going every September."[40]

It was hurricanes, however, the "equinoxial storms" so vicious in the Gulf, that taxed every aspect of Mercer strength and technology. "September 16, 1875. 1AM. All hands got up. The wind blowing fearful from north-north-west. The tide up to the gate . . . water over waist deep. . . . 4 AM. Barometer 29.4. The wind ever blowing from north by west making things rattle with rain. . . . Tide rising slowly. . . . 6 AM. Barometer 29.3½. The storm increases. . . . Several chickens had the life blown out of them. . . . 5 PM. Barometer 29.1¾. Winds north-north-east screaming . . . with plenty of rain. . . . 7 PM. Barometer 29.1½. . . . Rain falling in earnest. . . . such a storm in the Cove as never was."[41]

The damage never lessened. "St. Josephs Island . . . had a severe time during the blow. They [have] abandoned their houses and taken to the sand hills." Nor did time diminish nature's blasts. "A . . . northeast wind began . . . Thursday evening, and . . . shifted to the northwest when the wind became terrible," a Corpus Christi resident reported eleven years later. "From 5 to 7 o'clock Friday morning the wind was at its height. Fences, trees, sheds, wind mills, arbors, gutters, out-houses . . . were . . . falling in all portions of the city." This was the storm that ultimately obliterated Indianola. Virtually the entire town cratered in the gale, a survivor recalled; "fire destroyed the signal corps officers and crew . . . and nearly every sailing craft on the coast" was tossed ashore.[42]

Like wind, the sea also destroyed, as it did Sabine Pass two months later. "The hurricane . . . was maintained at high point by the mysterious waters behind. . . . [They] began to invade the town from the gulf . . . and rose with unexpected rapidity. . . . Only six houses remain standing in the town and they are off their foundation," a reporter wrote. Over one hundred persons were gone.[43]

Such ocean-borne attacks led even the *Philadelphia Times* to suggest "very clearly that the entire Gulf Coast of Texas and even Louisiana lies too low to furnish safe sites for any considerable towns."[44] But those were inlander fears. People who thrived on the coast continued to embrace their islands and bays—and inadvertently strip them.

CHAPTER 9

Harvesters

It was not done hostilely, or even consciously. If questioned, any South Texan would have referred to the Old Testament and its directive letting humanity "have dominion over the fish of the sea, and over the fowl of the air, and over the cattle." But few questioned that concept; even the Mercers, as close to nature as any Coastal Bender of the day, hooked baby turtles and harnessed porpoises for rides: "first time anyone here used a fish for a tug boat."[1]

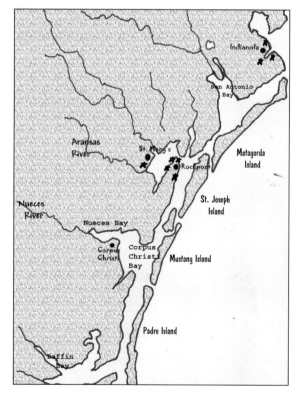

Fish and turtle processing plants of the late 1800s. Based roughly on Robin Doughty, "Sea Turtles in Texas."

Oyster beds around Aransas Bay. Courtesy of Texas Maritime Museum.

Flora and fauna of the barriers seemed to exist for no other reason than to be reaped, oysters most especially. Not the animals themselves. Until the 1880s, those were primarily a local product and best when served fresh, as one advertiser promised: "Stow and Company are now . . . catching oysters . . . and will serve all who desire it, at their residences." But it was the bays within which these shellfish flourished—the warm nutritious waters that cushioned and nourished their larvae—and the ever-growing reefs upon which the growing spat attached themselves that became targets.[2] City leaders hopeful of improving roads were among the first to subsidize the razing of oyster beds. "The Mayor called the attention of the Board to the necessity of shelling the streets," a *Galveston Bulletin* reporter wrote in 1886. "He . . . recommended that the Street Committee be authorized to invite proposals for furnishing . . . 100,000 barrels of Reef Shells." Others joined the movement and added suggestions: "Put in a layer of oyster shells with fine shells on top. This will hold sand down and make a fine street." In time, mudshell, as it was to be known, became ubiquitous on coastal roadways. "Gravel has been replaced by shells," one writer remarked, "and the change indicates that one is not far from salt water."[3]

Road construction was indicative enough of coastal harvesting. But landsmen went one step farther when they rediscovered shellcrete. It had been developed earlier by Spanish settlers, who combined clay and lime (burned shell ash) with crushed oyster shells to make adobe-like bricks; late nineteenth-century builders added portland cement to the mix. More solidly compacted blocks composed of lime, crushed reef shells, water, clay, and cement began to be manufactured. By the start of the next century, new shellcrete structures were ringing bays next to old residences like Fulton Mansion near Rockport and the Old Centennial House in Corpus Christi. Adding to the advancements was the recently invented steam-powered dredge. Now contractors no longer had to rely on shovels and wheelbarrows to fill their orders. Power-driven pails could thrust deep into subtidal depths and scoop up generation after generation of oysters, their shells, and their reef-bred cohabitants: sponges, tube worms, shrimp, crabs, mussels, slipper snails, and tiny fishes. It was no slight incursion.[4]

Nor were the seine nets hauled over shallow parts of Galveston and Matagorda Bays in the 1880s. Woven with small mesh to capture white shrimp feeding in the waters, they brought in over one hundred pounds a day for local fishermen, in season.[5] It was cast nets, however, that combed Corpus Christi and Aransas Bays at that time. Round, small webbed, and weighted by sinkers, the devices were little different from those fashioned by the Karankawas centuries earlier. But handled by men and boys permanent to the area, they cut recurrent swaths through muddy habitats and raked up half-grown shrimp by the pound. Prized for their tenderness and accessibility, the crustaceans, like oysters, were usually sold fresh, "brought to market and placed on the stalls by the fishermen and . . . bought up . . . by . . . speculators, who sell them to the public."[6]

No harvesting, however—neither that of shells nor oysters nor shrimp—matched the taking of green sea turtles. Tiny little beings, gnawing doggedly out of barrier-island nests and scuttling desperately toward the sea, they thrived once they hit the waves of Padre and Mustang Isles. Diving, swimming, scrounging for aquatic insects and minuscule worms, they grew into adolescence, their eyes protected by a single set of scales and their shells bony, plated, and occasionally green. It was their jaws that distinguished them, however—serrated and strong, they enabled emerging adults to subsist on sea grasses and herbs alone, eschewing the carnivore diet common to other such reptiles.[7]

Until the beef-and-tallow industry came to the bays. Incessantly harangued by health advocates for their fly-infested waste mounds, proprietors began shoving carcass and bone remains into the water. Already clouded with town sewage and crop runoffs, the estuaries began to resemble rancid smorgasbords, offering soft, easily digested carrion to marine life more used to sea grasses. Young turtles were among those enticed, and their bulk, as they grew and thrived in the inlets and lagoons, became noticeable. "They would weigh three or four hundred pounds," an observer commented, a significant increase from their standard two to three hundred.[8]

And in an irony as sad as it was deadly, turtles grew larger while longhorns grew rarer. Summer droughts and winter blizzards contributed to the herds' decrease, as did a growing demand for better stock. Entrepreneurs like King turned to developing their own breeds, but packery owners seemed doomed—until they looked into the bays.[9]

Turtle trussed, awaiting shipment. Courtesy of Port Aransas Museum.

Soon, twined nets up to one hundred feet long and ten feet deep materialized, their meshes almost sixteen inches square. Designed to entangle, they draped almost invisibly within the channels of Aransas Bay to snare cruising turtles. Two or three captured a day was judged to be satisfactory, one chronicler reported, "although it was possible to secure as many as twenty animals daily." Trapped by head or flipper and raised landward, many were sold immediately on the open market, as were oysters and shrimp. Others, however, especially those meant to be processed, were placed in open pools. "We used to watch them come up and put their heads out of water to blow, as we called the act of breathing," an onlooker recalled. The "crawls," pools at least one hundred feet in diameter, were buttressed by wooden stakes strung together to provide strength. Within them, the captured turtles bobbed and swam until the pens were full. Then, those that were to be kept alive until transferred elsewhere were "turned on their backs and trussed for shipment." The rest "were hauled out by block and tackle and butchered on the wharf" by former packery employees, who sent their flesh to the numerous canning companies that had sprung up along the coast. By the 1880s over four hundred thousand pounds of green sea turtles had been captured along Aransas, 80 percent of the entire number taken in Texas that season.[10]

As the adolescent and mature turtles decreased, so also did their nests and their hatchlings and their habitats. Sea-grass beds, already affected by waste and effluent, lost the winnowing done by their reptilian dwellers. Gravid females no longer crept upon Padre and Mustang shores to lay and cover fertilized eggs. Beach sands on the isles no longer bore body pits from which hatchlings emerged. As the decade flowed onward, green sea turtles of the Coastal Bend began to disappear; by 1890, "the Texas turtle industry had been virtually wiped out"—as had most of the turtles.[11]

No less affected were the Bible's "fowls of the air." Once apparel manufacturers convinced society matrons that only elaborate, feather-strewn hats could be stylish, skies and marshes of the islands became shooting ranges. Gray herons, white pelicans, blue jays, and yellowhammers were brought down, seagull and tern rookeries torn apart, geese and crane nests decimated. Men who netted turtles during the warm months and oysters in the winter sported rifles and shotguns during the spring and, like the soldiers in Taylor's troops, bragged of their count: "crane, shear-water,

Texas scow sloop, loaded with fish. Courtesy of Texas Maritime Museum.

pelican, snipe, duck, dove, mockingbird, etc." But unlike the members of the occupation army, these fishers and hunters were permanent and added their own sloops, catboats, and schooner-rigged scows to the traffic on the bays.[12]

For bounty was there, plentiful if not specialized as shrimp, turtles, and oysters had been. Freshwater and saltwater fish, so available to the Mercers on Mustang Isle, still populated channels and estuaries and, by midcentury, had attracted far more commercially oriented entrepreneurs. Some fleeing the equinoctial storms up the coast, others hailing from southern

Europe, and a few migrating north from Mexico, the newcomers helped push finfish yield to the top of the state fishing industry. Soon they were crisscrossing Aransas and Corpus Christi Bays, thrusting huge, weighted seines—six hundred feet long with mesh one inch square—down the sides of their skiffs and trolling the depths for booty.[13] By the end of 1880, their take of crabs, mullet, redfish, trout, and sheepshead had netted Texas close to $128,000.[14] The cornucopia of the barriers seemed boundless.

CHAPTER 10

Hustling with Hope

As did the promise of their waterways. Those of Galveston had been under innovation since 1850, when private interests cut an eight-mile-long channel down West Bay. Their success in maintaining and widening it led to an attempt to remove that ridge threatening Galveston Bay twenty years later. Rows of cedar poles, driven deep into the ocean, forced enough current through the inner and outer bars to scour off some sediment, but entry remained chancy. So the Island City asked the military "to put twenty feet of water on each of the two bars of our Harbor." No piddling twelve-foot clearance for this entrepôt; its leaders wanted the army to build jetties rivalling those of New Orleans.[1]

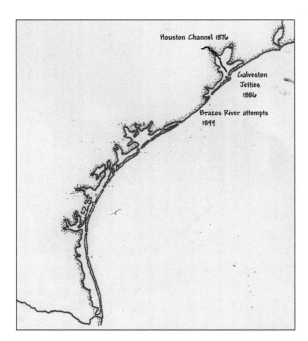

Active channel and jetty building in late 1800s. Based roughly on Francis E. Sargent and Robert R. Bottin Jr., *Case Histories of Corps Breakwater and Jetty Structures.*

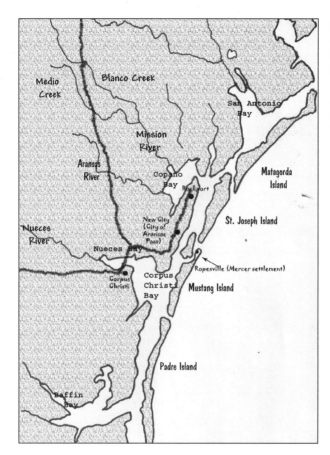

Route of San Antonio and Aransas Pass Railroad, 1886–1890. Based roughly on *Map of San Antonio and Aransas Pass Railway and Connections*, author's possession.

They had not taken the sea into account. In spite of a military design that filled vast hampers with dredged sand and cement and then anchored them off both sides of the pass, the baskets twisted and turned with the undertows until useless. Undismayed, the Army Corps of Engineers tried again, layering logs, brush, and rock onto mats extending twenty-two thousand feet off the northern tip of Galveston Island. By 1885 shipworms and burrowing creatures had destroyed what currents had not. It was not until the mid-1890s that bulwarks—built of sandstone riprap and granite—were anchored so successfully that they scoured the passage into Galveston Bay to a depth of twenty feet.[2]

But other ports had not been idle. Only Charles Morgan's refusal to rebuild piers and warehouses inland kept Matagorda Bay residents from redesigning their inlets, and by 1899, the Brazos River Channel and Dock Company had spent over $1 million trying to secure the flow of its tributary into the Gulf. Even inland Houston had obtained an outlet to the sea. Buffalo Bayou, situated on one of the state's most accessible waterways, "meandering . . . wide and deep enough for schooner or steam boat navigation," was seen as an economic asset by its earliest settlers even before the Revolution. By the 1850s, city and private interests had dredged a channel down the stream into the San Jacinto and out into Galveston Bay. By the late 1860s, Houston's port was using its own barge and lightering facilities to bypass Galveston's wharves entirely, and less than ten years later the Buffalo Bayou Ship Channel Company had gained Morgan's pledge to dig a canal for it through Galveston Bay into the sea.[3]

It was the Coastal Bend, however, that epitomized waterway alteration in the late 1800s. Even before the Civil War, go-getters—profit-driven entrepreneurs inspired by their nearness to the sea—had petitioned the state for permission to dig a six-foot-deep channel within Aransas Bay. This conduit, connecting with another trench in Corpus Christi Bay, carried with it all the financial "rights and privileges" of any privately funded project. At the same time, a similar measure, to dredge a canal along Laguna Madre from the northern edge of Padre Island to Brazos Santiago, was also passed. The demands of war scuttled the lagoon duct, but the Aransas Road Company successfully deepened the dugout[4] between Corpus Christi and Aransas Bays by two feet—until Colonel Lovenskiold jammed it with shellcrete in his desire to deny Kittredge entry.[5] Cleared by the Union commander days later, the scarred bay bottom lay silting until a similar dig in the 1870s reopened it. But it began to clog as well, emphasizing the shifting-seafloor quandary faced by lighter men in the bays and bar pilots in the pass.[6]

In the last decades of the century, however, Texas interests shifted from sea to ground shipping as wood-powered locomotives demonstrated the speed with which products could be propelled to inland markets. Charles Morgan again took charge, setting up a steamer-to-track monopoly with lines running from his Houston port east to New Orleans, north to New York, and west to San Francisco. The enterprise, called the Gulf, Western Texas, and Pacific Railroad, faltered in the mid-1870s but led to his acquisition of the Houston and Texas Central Railway three years later. This

The Belknap, first locomotive purchased by the San Antonio and Aransas Pass Railway in 1885. Author's personal collection.

system, running over 300 miles to West Texas and 160 miles to the northern and central parts of the state, gave the shipping magnate access to more markets from Buffalo Bayou wharves than he could ever have anticipated in his early days. It also intensified competition with the Mallory Shipping lines, which had monopolized railroads serving the Port of Galveston.[7] Morgan died before he could measure long-term results, but his vision of rail-and-water integration inspired other go-getters in the state.

One of them was one of the most persuasive pitchmen of that era, Colonel Uriah Lott. Like Morgan, the easterner had a dream, but his was more local: to move Coastal Bend bounty—trussed turtles, finfish and oysters, plumage and eggs, even cattle—from the sea by rail to markets all over Texas. His first attempt, the Corpus Christi, San Diego and Rio Grande Railway, was a bust, but it did convince two of South Texas's most respected entrepreneurs, Mifflin Kenedy and Richard King, to invest.[8] The old captain had gone before Lott's ultimate vision—a rail line connecting San Antonio with the Coastal Bend—was complete, but Kenedy protected his former partner's interests and those of the area. By March 1886 the system's roundhouse and machine shops were operating in the Alamo city, and less than two months later tracks to Beeville were laid. Followers watched eagerly as the San Antonio and Aransas Pass Railway moved south, and they cheered when its first engine crossed the reef road "to Corpus in perfect safety." As the inaugural train chugged into town a week later, citizens gathered on "the Old Camping Ground of Taylor and Grant" to rejoice.[9] Ranchers were happy for obvious reasons: "Beef cattle will be shipped from any point on the road through to St. Louis or Kansas City for $100 per car or to Chicago for $190. This is $5.00 cheaper than any rate ever given." But bay residents had even more cause to applaud: "Corpus Christi is . . . to have direct railroad connection with San Antonio," the *Caller* editor explained. "That will mean a great deal for us as the road will lead to deep water . . . a deep water outlet on the coast."[10]

Deep water: a vast, dredged-out anchorage where ships drawing every depth could be laden with goods unique to the area and where longshoremen would offload only the finest items from abroad. It was a goal desired by every self-respecting shoreside city west of the Mississippi, in the heady days following Reconstruction. But only New Orleans had been able to claim such a harbor before the Port of Galveston made its try. Within seven years of setting up a "Committee on Deep Water" in 1881, the city won federal support—and Army Corps of Engineers' expertise. By 1896, "the largest cargo ship in the world . . . drawing twenty-one feet" was moored at Galveston wharves. The "Queen of the Coast" had bested all competitors and claimed the throne.[11]

But Galveston could not change its location. In a fit of pique, a rival reminded the Island City of its unsafe position: "It is more exposed to the ravages of the Gulf storms" than any other community. In the meantime, its coastal counterparts continued to fight for ocean access. Houstonians turned to the government, the Brazos River Channel and Dock Company resumed work on its jetties, and the citizens of Aransas and Corpus Christi looked to their bays.[12]

CHAPTER 11

Hard Tracks and Haupt

And why should they not? The Coastal Bend's waterways were legendary and its rail access—the San Antonio and Aransas Pass to the northeast and a sister line, the Texas Mexican Railway, to the west—was enviable.[1] Connected to a deepwater harbor, these tracks could tie the borderlands of Mexico, the canyons of the Hill Country, and the bays and barriers of South Texas to the world.

Padre had been ready. A full year before the S.A.A.P.[2] extended its line to Aransas Bay, landowners on the isle had deeded to John Willett a good part of "the entire waterfront [of] that island," sixty miles of sand. His goal had

Jetty work on Aransas Pass, 1880–1896. Based roughly on Tom W. Stewart, "The History of the Aransas Pass Jetties."

been to sink a wharf of iron piles, build a breakwater, and thus create an "artificial harbor" within the lagoon. Optimism ran high. The *Fort Worth Gazette* deemed the site "the most feasible and economical point on the Texas coast [for] deep water," and even Henrietta Chamberlain King, the old captain's widow, thought about investing.[3]

Rockport, the erstwhile meat-packing center of Aransas Bay, had also been angling for deep water. Its survival of the hurricanes of the 1870s and 1880s was a selling point, as was its hookup with the S.A.A.P. in 1890.[4]

Corpus Christi remained available, its promoters quick to point out its "magnificent natural stand from which to view . . . the bay," its "seashore and suburban drives" underlain by shell, and its elevation, "more protected from storms than any other point on the Texas coast." "There should be a united effort made . . . to obtain access to deep water on the coast," an 1887 editorial admonished before Galveston got the bid, "Corpus Christi being . . . the best location."[5]

But it was a small area to Rockport's south that had most eagerly aspired to deep water. Acquired by a group of businessmen in the 1880s "with the intention of purchasing and dividing land" and tentatively christened New City, it aroused even more interest as the S.A.A.P. line approached. Within just a few years, its developers decided not only to connect their settlement with the railway but to extend from it their own tracks, crossing east over Aransas Bay and onto Harbor Isle. And there, between it and Mustang, they would dredge a port of such depth and accessibility it would become an international draw.[6]

But while investors flocked and bankers dreamed and petitioners lobbied Congress, roiling waves coursed new sands onto the isles' beaches, and currents culled their banks. Storms carved fresh inlets, and mudflats vanished. The distance between shorelines broadened while widths narrowed.[7] As organic in their own way as fish and sea-grass beds, the islands—and their pass—grew even as they diminished.

And therein lay the problem for deepwater dreamers, for how could one dig a harbor when its access, through which ships would steam, constantly shifted? Reasons for this varied, wind-driven streams being one factor, said one student of the coastal barriers, Colonel Elihu Ropes. The outgoing current, he posited, carries the channel through the pass and toward the southwest but is "seized . . . by the littoral current at the outer end" of the

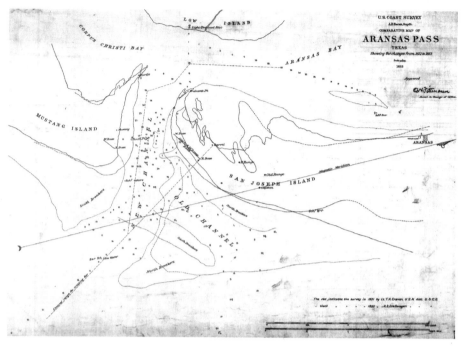

US Coast Survey map of the movement of Aransas Pass between 1851 and 1853. *U.S. Coast Survey Comparative Map of Aransas Pass Texas, 1853.* Courtesy of Port Aransas Museum.

isle and then pressed against the southern bank, which "it wears away constantly." Engineering instructor Major Cassius Gillette agreed, to a certain extent: "Generally . . . in the western part of the Gulf . . . the littoral drift, reaching an inlet, prolongs one side of it and, by causing erosion of the other, makes the inlet travel with the littoral drift."[8]

But there was another reason. "At certain seasons," he added, southerly winds blow until "the lagoons or bays are filled above their normal level. Then the wind changes almost instantly to a strong wind from the north. These are called 'Texas Northers.' They start at full speed, and the swollen waters of the lagoon are driven violently to the south, escaping to the ocean with great velocity . . . scouring a deep channel and eroding the south side of the inlet. This then is the cause of the movement."[9]

Regardless of cause, however, "the pass was moving bodily in a southwestward direction," O. H. Ernst of the Army Corps of Engineers reported, "the southerly shore being eroded and the northerly shore advancing by accretion. The rate of movement between 1868 and 1878 was about 260 feet per year." In addition, the widest part of the strait along St. Joseph's bay side was in flux, its waters expanding over thirty acres in ten years' time. But it was the unpredictable depth of Aransas Pass that warranted greatest concern. "It was 9 feet in 1852 and 1853, 7½ feet in 1868, 9½ feet in 1871, 7 feet in 1875, and 7 feet in 1878, all at mean low tide."[10] From any investor's point of view, such an erratic conformation held no reasonable chance of profit, unless it was stilled. So fixing the islands into place became the ultimate aim of Coastal Bend entrepreneurs.

Serious attempts had begun not long after Appomattox, when engineers erected a dike, filled with brush and rubble (riprap), along the north side of the pass on St. Joseph Isle. Mats designed to scour the bar were added later, but the entire structure failed, demolished by storms and sedimentation from the east.[11] Then the military, already chastened by its efforts in Galveston, tried. Called on by Congress to solidify the pass in 1880, the Army Corps of Engineers ordered Major Samuel M. Mansfield to the barriers. Operating on the same principle as earlier designers but starting from the opposite island, Mansfield planned the structure to direct sand-bearing currents away from Mustang into open waters. Consequently, he built his jetty off the eastern edge of the isle and extended it four thousand feet out, its foundation passing the old St. Mary's wreck and arcing northward at its tip. But Mansfield was hampered by insufficient resources, and the light rock he loaded upon his embankment held no sway against full ocean thrust. Moreover, teredinids, seaborne mollusk borers, gutted the felled trees and branches he had used as ballast. Despite slowing the pass's migration a significant amount, his jetty failed to deepen it. Within three years, the structure's farthest outcrop was sinking, sucked into currents it had failed to check.[12]

So the army came back, this time under the command of Ernst. The officer's 1886 mission was to confirm that the channel had deepened, the wind had ceased to erode, and the pass was immobile. He could not do so. "The work designed to deepen the channel over the bar has had no important effect . . . the sand fences and plantations of salt cedars . . . to stop erosion . . . by wind . . . have disappeared," and the northern tip of Mustang was still receding. Mansfield's efforts, the major dryly noted, "had not proved entirely successful."[13]

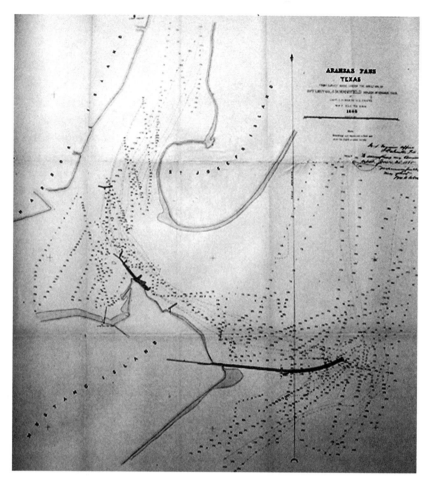

The Mansfield Jetty, as charted by Lieutenant Colonel Mansfield in 1885. *Aransas Pass Texas from Survey Made under the Direction of Bvt. Lieut. Col. S. M. Mansfield, May 21st to 25th, 1885.* Courtesy of Port Aransas Museum.

Ernst took on the job. First, the part of the 1880 jetty still above water was to be recurved south, and thus connect "by the shortest line to deep water." A second bulwark was then to be built on St. Joseph Island, about two thousand feet away from its sister structure, both strengthened with rail-delivered limestone and granite. Together they would offset the power of "millions of cubic feet of water flowing at velocities . . . from 6 inches to 6 feet or more per second . . . and obstruct this flow just enough to excavate the channel . . . to . . . a depth over the bar of not less than 20 feet." The width within the passageway between St. Joseph and Harbor was to be adjusted as well, the major added, for changes were in store.

"During the last year, a railroad has been constructed from San Antonio southward, and is now open to traffic to Corpus Christi Bay. Its terminus is to be . . . on Harbor Island, where wharves for sea-going vessels will be established. . . . [This] makes the requirements of this harbor wholly different from what there were when only . . . vessels of light draught were to be accommodated. . . . The more important object . . . now is to admit ocean-going ships drawing from 18 to 20 feet." Major O. H. Ernst had transcended the limits of army engineering and become an unofficial advocate for deep water.[14]

But first, to stop "the southward movement of the pass," he sent his assistant J. E. Savage to blanket the far edge of Mustang Island, from its "high-water mark to the bottom of the channel," with rubble, gravel, and stone.[15] And for the first time since their ascent eons earlier, the mobile sands of Mustang's tip froze. Twenty-seven thousand feet of beach, long awash with wave-driven sea sediment and shells, found itself compressed under twenty-three thousand cubic yards of rock. The work was done carefully. Using sextants, range markers, and bottom-fastened tag lines, Savage moved a barge to the designated site, unloaded eighty-five feet of rubble, then steered another to its place. That one emptied, another came and then another and another. Finally, almost two years later, he had anchored a half mile of the island's tip in stone. For further security, the officer built a small wall, extended it one thousand feet, constructed three spurs on it, and fixed it to the shoreline.[16] Aransas Pass was now partially riprapped, encased, and hedged—could a world-renowned ship harbor be much further off?

Surely the avid entrepreneurs selling property in the little settlement south of Rockport believed one was at hand. Already street plans had been laid for the New City, rail lines were drawn, and advertisements appeared, showing schooners and steamers cruising freely through the pass into widened waters. Savage's embanked area of Mustang even appeared to be holding steady. Despite vicious tides and brutal northers, "the interstices between [its] rocks had filled," he reported proudly, "a thick coat of seagrass covered the riprap, and drifts of shell and sand were found parallel to the shore-line." All seemed set for Ernst's next phase, the erection of twin jetties—until Congress directed the Corps of Engineers farther north. Galveston had officially been designated Texas's first deepwater port, so no more money remained "for the improvement of Aransas Pass."[17]

But money could be raised for a brand-new pass, insisted Elihu Ropes, fresh from coastal investigations and full of excitement. With private financing, the colonel began to create his own passageway, dredging a channel across Mustang Island and planning, when it entered the Gulf, to erect two wooden jetties twenty feet long. One would be sunken, a "drowned jetty," he explained, where the waters coursing over it would carve a trough so deep "it will open the Pass . . . to vessels drawing from twenty to twenty-four feet of water." In time, a more solid bulwark would be erected, a port dredged, and a railroad built, to extend south from the new port to the lagoon. That it would cross, carrying goods lightered from heavier ships to deposit at mainland terminals. In the meantime, shallower-drafted vessels could sail directly through the new pass to wharves on Corpus Christi Bay. It was a bold and innovative idea, sanctioned by Ropes's in-house engineer, H. C. Ripley, and especially attractive to investors frustrated by the government's abandonment of an Aransas Pass port.[18]

Even as Ripley prepared his report, however, investors had begun to question the stability of Ropes's financial arrangements. The next two years saw creditors attacking, and even before the depression of 1893 officially set in, the colonel had fled the Coastal Bend, his dredger stuck in sands briskly resilting his partially dug channel.[19]

Backers in Corpus Christi, the town most featured in Ropes's scheme, were furious. But even more downhearted were supporters who lived in the tiny hamlet off the tip of Mustang Island. Long inhabited—its earliest families included the Mercers—the enclave boasted the pilothouse originally part of El Mar Rancho, the lifesaving station first manned by John Mercer, a post office facility, a general store, a resort inn, and fewer than two hundred souls.[20] With his plans, Elihu Ropes had galvanized the settlement, envisioning for them an influx of laborers, merchandisers, land agents, and home builders, all seeking opportunity at the Mustang Island port. With his flight, they deflated. Almost bereft, they removed his name from their town and, as a later editor wrote, resumed, in their "unassuming manner, [their] own affairs and left the big world outside to do likewise."[21]

Also shaken by the loss of government support for a navigable Aransas Pass, but not quite as credulous, the people of New City chose an option different from that of their bay neighbors. In 1890, Russell Harrison and Tom Wheeler, one a president's son and the other a former lieutenant governor of Texas, created their own company, its goal to dig the harbor Ernst had proposed three years earlier. Their engineers would create a jetty, extend a channel from the pass along St. Joseph, and then deep-dredge a port off Harbor Isle. From there, as earlier planned, the S.A.A.P. rail system, its owners strong supporters of a deepwater port, would transport goods from ships to trains to the mainland and thence across Texas, Mexico, and the West. The entrepreneurs were so confident that in 1892 they even renamed their New City after the embodiment of their success: Aransas Pass.[22]

To no avail. Within a year of its undertaking, the J. P. Nelson jetty, extending over one thousand feet off Mustang Isle and only lightly burdened with rock, cratered onto the sea bottom.[23]

Their hopes heading in the same direction—S.A.A.P. support had already disappeared with the financial unsteadiness of the 1890s—Harrison and Wheeler hired two of the most authoritative engineers in civilian service to save their deepwater port: H. C. Ripley and Lewis M. Haupt. Ripley was already a familiar sight in the Coastal Bend. Earlier days with the Corps of Engineers, mapping signals and benchmarks off St. Joseph Island, preceded his work with Ropes. Haupt had served with the army as well but by 1890 was a professor of civil engineering at the University of Pennsylvania. It was his observations of currents in lakes, rivers, and Atlantic harbors that led to the design that he promised would secure the pass virtually forever: a single reaction breakwater.[24]

It was this single reaction breakwater, partially implemented but roundly critiqued, that demonstrated publicly the ever-changing phenomena of the Coastal Bend. For in examining the structure's eventual failure, investigators rediscovered the region: its prairieland, for example, "the broad . . . slightly rolling sandy . . . terrace . . . that borders the shore." Identified as "emerged shallow sea bottom," it had originally been "a portion of the sea's continental shelf," one authority noted. It was the still-submerged part of that shelf, he went on, that powered the islands' breakers. When "the waves from the deeper seas attain the shallow water next the shore, they begin . . . sweeping sands in. . . . As the friction of the bottom of the wave . . . increases, the upper part shoots forward . . . flies clear beyond its base . . . and breaks upon the shore."[25]

Another phenomenon fairly unique to the proposed venue was the barriers' littoral drifts. Unlike most currents propelled by the Gulf Stream, those alongshore of the isles experienced little push from it, more like a "gentle" nudge. In addition, tidal movement within the lagoons and bay was slight. Where other coastal tides rose and fell several feet daily, those in that part of the Gulf waxed and waned only slightly. Their primary force, an engineer concluded—pushing sediment from the southeast some months and detritus from the northeast during others—came from the wind, diagonal, hard driving, and "acting through the waves and currents it produces." That such wind, in certain circumstances, could turn into "Texas northers" and twist whole inlets awry was one more distinction of the Coastal Bend.[26]

In spite of these singularities, or possibly because of them, Ridley and Haupt guaranteed their breakwater's efficacy, the latter promising its construction would "unquestioningly result in securing navigable depth over the bar at 15 feet for the first part of the work and 20 feet for the second." His design was simple, its final form extending into a bulwark of stone 6,200 feet long, with "a top width of 10 feet, rising to a height of 3 feet above . . . mean low water." Its slope would be rather steep, "1½ horizontal to 1 vertical," and its base would vary, depending on seabed depths, from 40 to 70 feet. The key to Haupt's plan, besides its relatively low cost, was its uniqueness: fixed to neither Mustang nor St. Joseph Island but to the obstructing bar itself, the structure was to curve inward from the channel for its first approximately 2,000 feet, curve outward another 2,000 feet, then inward again for its remaining length. The shape, Haupt felt, would preserve the full force of waters streaming through the pass and compress them to such density that the ensuing downdraft (reaction) would flush their sediment outward and across the gap, deeply gouging the floor at the same time.[27]

The first phase of installment would lay the central part of the bulwark and a significant section of one extension. Even just these, assisted by seasonal northers, Haupt assured, would scour a consistent fifteen feet throughout the pass. The remaining five feet of depth would come with the rest of the structure, and since, once excavated, the channel need only be "maintained by the natural tidal courses from the bays, the cost of maintenance . . . would be . . . minimum." Thus promised efficiency as well as economy, the owners of the Aransas Pass Harbor Company accepted Haupt's plan, purchased 25,500 square yards of brush mattresses and 57,700 tons of stone, lay three and a half miles of tracks across Aransas Bay, and in August 1895 set about constructing the only single reaction jetty they would ever need for a deepwater harbor.[28]

The effect below the surface was varied. As contractors for the railroad line pile-drove creosoted timbers deep into the bay bottom, they pierced through myriad habitats, burying millions of simple-celled organisms, bivalve clams, and other benthic beings. Additional trestles stretching across tidal and mud flats destroyed insect and algae colonies and dislodged waterfowl. The makeshift wharf at their terminus, the Morris and Cummings Cut—that old channel between Aransas and Corpus Christi Bays—ravaged a bay bottom already scored from previous dredgings, while barges, onto which the railcar's cargo was offloaded, gashed sea-grass meadows and marshes as they were tugged toward the pass. At the same time, the timbers piercing the floor offered fresh shelter to barnacles and sea snails. Teredinids latched onto their bases, and minute organisms burrowed into newly excavated troughs. Tiny crabs and baby shrimp clustered close to the emerging jetty, and terns and sea gulls flocked to feed on neighboring mullet and small drum.[29]

But it was the rocks—huge, crudely hewn blocks heaved from the barges by cranes and lowered below the surface—that had the greatest impact on Aransas Pass. Stowed three feet high on a wood-filled mattress stretching sixteen yards wide, sandstone chunks compressed the ocean floor even more densely as smaller boulders were piled atop them. Reaching the surface at mean high-tide level, the stones were then cap-rocked to an overall height of twelve feet. About 1,500 feet from the pass itself, the structure, secured to one of its most obstructive bars, curved eastward, crushing downward over 5,000 feet of ocean floor.[30]

Its first phase of installment completed as promised, Haupt's breakwater dominated the Coastal Bend. Hydrologically, it displaced waves, circumvented littoral currents, and hollowed out gulches. Culturally, it made the pass a target for newspaper editors and realty agents. Scientifically, it dominated engineering journals and conference gatherings. Politically, it gave new life to forgotten party bosses. Economically, however, it failed.

Maps revealed all. One drawn by H. C. Ripley in 1895, on the eve of construction, showed depths of 6 feet about midway in the pass, varying

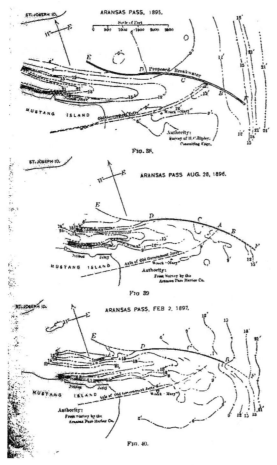

Charts revealing the failure of Haupt's single reaction breakwater between 1895 and 1897. *Transactions of the American Society of Civil Engineers*, vol. 54, part A.

to depths of 12, 15, 21, 24, and 12 feet closest to Mustang Island. Another done by the Aransas Pass Harbor Company a year later, with most of the breakwater's central bulwark done, showed water levels of 6 feet midgap and, from there to Mustang, depths of 12, 15, 24, 12, and 9 feet. A map of February 1897, done seven months afterward by the same company, measured a floor that carried 6 feet of water at midpoint, 15 and 21 feet closer to the isle, and finally 6 feet at Mustang's tip.[31] Other charts drawn during the same period showed soundings ranging from 15 feet to 6 feet across the bar. Fluctuating but irrefutable, the numbers indicated that water levels in the pass were still dangerously erratic.[32] Haupt's first installment guarantee of a consistent 15-foot navigable passageway had not materialized.

CHAPTER 12

Hydrodynamics and Dynamite

But that conclusion was hard to discern in the flurry of Harbor Company publicity: its promises of "rich cheap lands . . . to northern agriculturists," its assurances that "water over the bar . . . is over sixteen feet," and its promotion of Aransas Pass city's first Green Turtle Barbeque and Town Lot Sale.[1] Some clues, however, were evident. In early 1896, directors requested a three-year extension of the breakwater completion deadline, and in February the Town Lot Sale fell so short of anticipated revenue it was abruptly shut down. Sharp-eyed reporters also noted chicanery. On the excursion barge pulled over the bar during the sale, a writer from the *Texas Sun* observed depth measurements being taken. "The man who took them . . . took the first few soundings inside the bar and called 'eighteen feet'; while going over the bar his arm got tired, his eye glared, and he passed from labor; when the other side of the bar was reached,

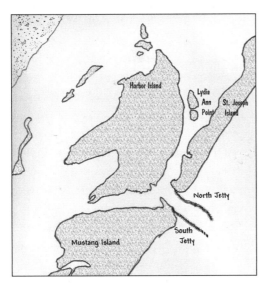

North and south jetties on Aransas Pass, 1909. Based roughly on Tom W. Stewart, "The History of the Aransas Pass Jetties."

soundings were again commenced and 'eighteen feet' was announced. There was not a man on that barge," the newsman noted drily, "who will not go home and tell his neighbors that there are eighteen feet of water over Aransas Bar and really think he is telling the truth." Longtime businessmen like John Willette, who had earlier entertained dreams of deep water near Padre Island, were also suspicious. "I assert in plain English there is not fourteen feet of water in a continuous channel across Aransas Bar and has not been for many years. Nine feet scant at mean low water is all that can be claimed or has been on the bar since the commencement to build the rock mound, to date."[2]

By September, the truth was evident: "Another attempt to secure deep water on this part of the Texas coast has fizzled." Contractors hired by the Aransas Pass Harbor and Improvement Company had quit. An editor summarized the situation: "When the work of deepening the channel at Aransas Pass was begun, over twelve months ago, there was from nine to eleven feet of water on the bar. Now it is estimated from four to seven feet, thus leaving several fine bays without an outlet to the sea." He finished bluntly, "The people are sick."[3]

But not the company. Within weeks its directors negotiated an agreement with a new contractor, Colonel H. C. Goodyear, under which "deep water has been guaranteed at Aransas Pass for the sum of $500,000." Sanctioned by Haupt and Ripley, the colonel promised to resume work within sixty days—and immediately ordered twenty-five thousand pounds of dynamite to be shipped to the pass.[4]

Dynamiting an obstructive bar to achieve long-term channel passage had been tried only once before, along the Atlantic coast off Brunswick, Georgia. Its use in the western Gulf, with its variant tides and offshore drifts, would be a challenge. Goodyear confronted it boldly. On the last Tuesday of November, his engineer set off a series of explosions, every ten minutes detonating a twenty-five-pound charge embedded in the bar impeding Aransas Pass. The results were immediate and positive: "a decided deepening . . . everywhere along the line" and the shoal penetrable for a good twenty-four hours. Continued blasts would make the bar so pliable that Haupt's single reaction breakwater could be completed and his guarantee of a full twenty-foot clearage ensured. Euphoric, directors of the Aransas Pass Harbor and Improvement Company began a full-fronted assault on the state for purchase rights to Harbor Island and its environs. Five months later, those same directors were in disarray. Local officials had abandoned them, the pass was still obstructed, and Colonel Goodyear's dynamiting had failed.[5]

Reasons varied, many rooted in the ocean floor itself. Sand based, parts of it secreted hard claylike substances while its shoals cemented shells, more sand, and ocean gravel into an almost impenetrable state. Slackening such nonerosive strata demanded far more than synchronized dynamite blasts, and their observed mobility after the explosive shocks was a result more of tidal flow than of artificial stimulus. "Sand on ocean bars . . . is . . . 'alive' at flood tide and hard at ebb," one military engineer noted. Because tides occurred more often "on the Atlantic coast, the sand, without any dynamiting, is 'alive' for six hours and then hard for six hours. . . . On the Gulf coast the periods [of "aliveness"] are twelve hours," sometimes extending even longer, depending on the movement of the tides. Any attempt to probe the bar during the long period of flood tide would reveal a certain ease in the structure, just as a poke during ebb tide would rebound. Because Goodyear did not take this Gulf-specific trait into account, the reactions recorded as evidence that his incendiaries significantly loosened shoals proved little. Even more telling, however, were the actions of the colonel himself. By the end of February he had left Texas. Upon returning in June, Goodyear acknowledged "a temporary suspension of work" but hoped to "overcome with ease all [difficulties] that may present themselves in connection with the Aransas Pass." It was too late. Government inspectors were already hastening to the scene.[6]

For three months, officers of the Army Corps of Engineers studied the Coastal Bend. By December their conclusion became public: not only had the Aransas Pass Harbor and Improvement Company failed to significantly open the pass, but its efforts were so faulty, necessity demanded that Congress retake control of operations. Federal acceptance of the task and commencement of work were slowed by war, incompetence, and party politics. It was not until June 1903 that a final contract to open the pass was granted—to H. C. Ripley. "Haupt is an eminent engineering authority," a Galveston editor commented. "With Mr. Ripley as his representative . . . the Haupt plan of one jetty system . . . will be given a thorough test . . . at Aransas Pass."[7]

It flunked. A year later, experts reported the "curved jetty not a success." "Results were not what were expected," the *Galveston News* headlined; "the channel is not deepening at the pass." Lewis Haupt seethed. In a scathing letter to the *San Antonio Progress*, he lay blame on the Old Government Jetty "lying across the channel, which prevented further deepening." The Army Corps of Engineers had purposely allowed the structure, also termed the Mansfield Jetty, "to remain, to restrict the scour," he maintained; it and rough waters in the bays had impeded his breakwater's effectiveness.[8]

Corps officers could do nothing about windswept bays, but they had much to say regarding his charge "that a navigable channel could not be maintained on account of the Old Government Jetty." Years earlier, they responded in the *News*, thirteen thousand pounds of dynamite were used to blow up the farthest extension of the old jetty. Any part that could have impeded Haupt's breakwater had already been shattered into oblivion.[9]

As were innumerable finfish, crustaceans, mollusks, and snails. In the decades that had passed since Major Mansfield had cautiously erected his structure, sea beings had colonized its cavities. Murexes skewered prey attached to its surface, pinfish devoured tinier shellfish, and red snappers fed on crabs. In its hidden spaces, female octopuses had laid eggs, and along its frame, algae and coral clustered with the sea worms that had originally gutted its mats. Like the shrimp, jellyfish, mullet, mackerel, and junefish that had been scavenging the Aransas Pass sandbar on the day Goodyear detonated his charges, Old Government Jetty wildlife died by the thousands, fish bladders punctured, shellfish fragmented, larvae decimated, and octopuses pulverized. With its seafloor cratered and covered with thousands of fish so immobilized they could neither swim nor breathe, the longtime habitat had been shock-waved into smithereens.[10]

As had Haupt's guarantees. To further substantiate the professor's failure, the corps quoted his initial promise that construction of the single reaction breakwater "as designed will unquestionably result in securing navigable depths over the bar of 15 feet for the first part of the work and 20 feet for the second part." Then they cited an Aransas Pass Harbor and Improvement Company chart that, in 1895, "showed a channel on an irregular line to a depth of 13 feet," then a year later, "after work done under the Haupt recommendation," an irregular channel of "only 10 to 11 feet." Engineers stated the obvious ("an irregular channel is far from being a navigable channel"), then defined the jagged passageway. It was "commonly called

... a thalweg" line: too inconsistent in depth ("one point will show twenty feet of water, while within a few feet only a depth of 11 feet is obtained"), too convoluted for use ("a vessel cannot follow . . . its windings in and out"), and too close to the barrier ("its . . . proximity to the jetty" renders it "not serviceable").[11]

Finally, the army brought out Captain Edward T. Mercer, "well-known pilot at Aransas Pass," to complete its case. Robert Mercer's grandson and Ned's middle child, Tom, as he was generally known, was in his late twenties when he reported to the corps. But he was already an accredited navigator with up-to-date tools, including mechanized devices to measure passage depths. His testimony, therefore, was vital—and excruciatingly detailed. "With light east winds and smooth seas, under favorable conditions and good tide, I can at present with the existing curved jetty channel bring in over Aransas bar the following drafts:

vessel 104 feet long drawing 10 ½ feet,
vessel 200 feet long drawing 9 feet,
vessel 300 feet long drawing 8 feet,
vessel 400 feet long drawing 6 feet."

The reason for deep-draft boats having to be increasingly smaller to access the pass was evident, Mercer added. "To follow the curve of the jetty a vessel 100 feet long must keep at least 200 feet from the jetty; a vessel 200 feet long must keep at least 300 feet from the jetty. Anything over 200 feet could not follow the curve of the jetty." An alternative opinion, that of site engineer F. Oppikoper, seconded Mercer's observation. "Pilots can bring over the bar, in ordinary high tide and with fair weather, vessels 100 feet long, drawing 10 feet. Vessels longer than 150 feet would have to draw less water." Despite the tons of riprap, granite, and limestone Ripley had laid, a deep enough, wide enough passageway still did not exist; rather, forces created by Haupt's breakwater seemed only to have scoured dangerously close to the structure itself.[12] The pass remained unyielding.

Until 1905. In August of that year, Darragh Brothers of Granite Mountain, Texas, commenced quarrying thirty-five thousand tons of rock. Within a year, those boulders had been transferred to the Coastal Bend, tugged through Aransas Pass, and then deposited on the seafloor. In November 1906 former lieutenant governor T. B. Wheeler could confidently state that

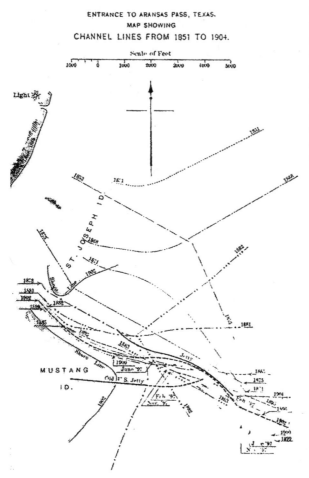

Chart showing continued failure of Haupt's single reaction breakwater to gouge Aransas Pass effectively. *Entrance to Aransas Pass, Texas. Map Showing Channel Line from 1851 to 1904.* Trans- actions of the American Society of Civil Engineers, vol. 54, part A.

one of the unique characteristics of Haupt's breakwater, the gap between the structure and St. Joseph's Island, was being filled. But "the channel is still too tortuous and narrow for practical use," he admitted, and he urged government engineers to open the pass.[13]

He was not alone. Other Texas politicians had been pushing for the waterway's improvement, including John Nance Garner of Uvalde, Walter Gresham of Galveston, and Richard Burges of El Paso.[14] But only one had the effrontery to suggest additional work—a passageway cutting west from the pass, flowing directly between Mustang and Harbor Islands, and

Hydrodynamics and Dynamite

heading straight to Corpus Christi. He was Rudolph Kleberg, member of one of Texas's leading families and early advocate of what was to become the greatest waterworks program in the South—an intercoastal canal.

All Kleberg had done was petition Congress to accept a privately constructed channel "to and through Turtle Cove," along Mustang Island, and "into Corpus Christi bay."[15] The idea was to ease access to the town's port without vessels having to sail north between Harbor and St. Joseph Islands before crossing through the old Morris and Cummings Cut and heading southwest. Behind his efforts, though, were the same leaders determined to bring deep water to the city—and behind them was King Ranch, in the person of Robert Kleberg.

Taking the reins of the company after the death of the old captain, Kleberg had continued to promote the interests of the Coastal Bend. Included in these were tick eradication methods he had developed to preserve cattle, irrigation attempts designed to supersede native climate, and land-selling excursions offered to disgruntled midwesterners. But foremost in the manager's mind was the need for international trade, and he saw that as achievable only with a deepwater port that was nearby and accessible. Consequently, he and his colleagues had stayed abreast of every development improving Aransas Pass. But as fervently as he desired a first-class harbor for his city, Kleberg wanted it connected to markets both on the Gulf and inland. Building such an extensive canal had existed "in conceptual form only" for years. But older brother Rudolph's proposal made a start, as did the efforts of a new lobbying group from Victoria, Texas. By March 1905, a study to create "a navigable channel to Corpus Christi through Turtle Cove" was included in a Rivers and Harbors appropriation bill, along with "the examination of an inland waterway from the Rio Grande in Texas to the Mississippi River at Donaldsonville, Louisiana."[16] A Gulf-based intracoastal canal finally had the interest of Congress; now all that was needed was a traversable pass.

And the Army Corps of Engineers was working on that. Granting a contract to D. M. Picton of Rockport, it authorized a final renovation of the old single reaction breakwater into a curved but solid jetty, extending out over six thousand feet. First securing rocks ranging from small riprap to seven-ton crest stones, Picton loaded them on railcars, propelled those onto trestles, and then chugged them out to the Morris and Cummings Cut. Once there,

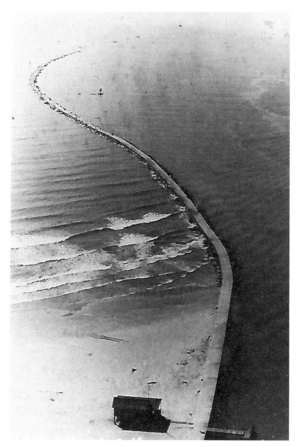

The north jetty on Aransas Pass, Haupt's curved breakwater now joined to the mainland of St. Joseph Island. Courtesy of Jim Moloney.

the contents were lowered onto barges and taken to the pass, where cranes deposited them on the seafloor, layering boulder upon boulder from the edge of St. Joseph Island to the revamped parts of Ripley's bulwark. Finally capped with cover stones of eighteen-ton weight, the new jetty boasted a crest at least ten feet wide, a height of six feet above mean low tide, and over one hundred thousand extra tons of rock to offset longshore drifts.[17]

But a single jetty, no matter how heavy, could not curtail currents well enough to adequately scour the pass—if nothing else, Haupt's single reaction breakwater seemed to have proven that. Included, therefore, in the $450,000 contract Picton signed with the federal government between 1907 and 1908 was a pledge to build a south jetty. And here he had free rein—no need to refurbish the old or match it to the new. Ironically, Picton's final design, a 7,385-foot-long bulwark stretching east from the tip of Mustang Island to a point 1,150 feet past the north jetty, was almost a mirror image of Major Ernst's plan of 1887. In both proposals, the jetties would be at least 1,500 feet apart and parallel to each other. The only difference was that Ernst had envisioned a new north jetty to act with a renovated Old Government Jetty, while Picton was building a new south jetty to work with his renovated Haupt Jetty. In any case, the twentieth-century contractor had his rock orders ready, and by 1908 over 250,000 tons of stone were on their way to Aransas Pass.[18]

This time, however, Picton and his associates improved on delivery. Rather than offload the rock onto barges to be pulled to the south part of the pass, he instead fitted the barges with the same dimension of track as that on the railway trestles. Thus, at the Morris and Cummings Cut, sandstone-bearing cars were chugged onto barges, tugged to the tip of Mustang Island, and then steered onto ocean-floored trestles. Upon these, the cars moved to the farthest end of the line where a huge crane awaited, there to carefully deposit the stones—larger ones on the sides and smaller rocks in the center—onto the seafloor of the pass. By the time the jetty reached its farthest limit, a solid center of layered riprap was sided by large sandstone boulders. Soon the breakwater, like its twin, had reached an elevation of

Stone-bearing barge hauling boulders to complete the south jetty. Courtesy of Jim Moloney.

Rocks preparatory to being lowered onto the jetty. Courtesy of Port Aransas Museum.

six feet above mean low tide; it had a crest at least ten feet in width; and it had two layers of sixteen-ton cover rocks.[19]

Most important, however, was its placement: one jetty, bearing a weight of 253,000 tons and extending south-southeast from Mustang Island, lay next to another virtually the same and parallel to it. Two extraordinarily embedded obstacles faced winds, currents, drifts, sands, "millions of cubic feet of water flowing . . . six feet . . . per second," and funneled them. By April 1909, the passageway was scouring itself "clean just as fast as the great walls are pushed out." Within months a new twenty-foot-deep, navigable channel had emerged—and with it, a virtually immovable Aransas Pass.[20]

PART IV. ENTERPRISE DAYS

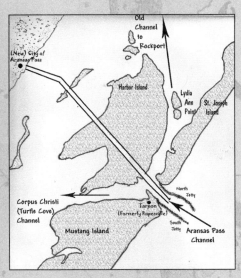

New channels, 1909–1917. Based roughly on *United States Coast and Geodetic Chart, 1918*, Port Aransas Museum.

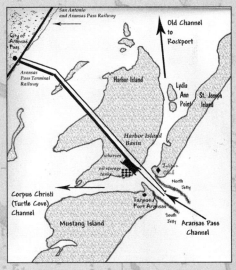

Harbor Island Basin. Based roughly on *Map of the City of Aransas Pass, Texas*, Texas General Land Office.

CHAPTER 13

Harbor Home

But not a silent one, at least not on its beach side. Most of the way down Mustang's twenty-mile length, birds still filled the air: dipping and playing the breezes, skimming sea surfaces, soaring the horizon, deep-diving the sea. And their sounds heralded them—squawks, caws, gulps, squeals, and cackles mingled with the clicks, whistles, and buzz of insects not yet devoured. The wind remained a constant murmur, almost invisible but for the salt it carried, while the surf added its own acoustics, a steady deepening thrum of oncoming waves crashing, then flowing seaward again.

Nor had silence enveloped the pass's swash, that rising ocean floor over which the waves rushed. Their breaks above the surface echoed downward and mingled with shipping sounds and mollusk movements; redfish and croakers added their own drummings to the waters.[1] Such cadences were not part of just Mustang Island; the eastward side of St. Joseph aired the

Birds in flight off Mustang Island. Courtesy of Jim Moloney.

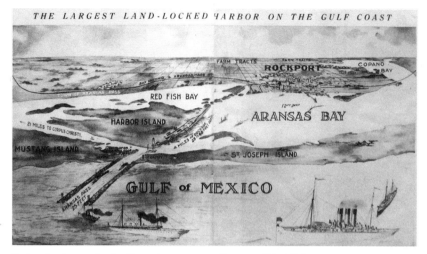

Rockport's ad, promoting itself as a deepwater port. Courtesy of Tom W. Stewart.

cries of gulls and the pounding of surf just as vibrantly as did the north end of Padre Island.

But it was among the back sides of the isles—those marshes and flats edging down to Turtle Cove on Mustang and Lydia Ann Channel off St. Joseph—that brasher noises reigned. For dredgers were there, diminishing barrier lands and recarving their bays.

The pass's opening brought them out. Rockport, one of the earliest to envision a deepwater basin off Harbor Island, had planned a channel and started a train line directly to the projected site almost immediately. Town fathers had even commissioned an artist to sketch the future port, with depths of twenty-five to forty feet, "four miles of dockage," and an S.A.A.P. rail connection "to Rockport and North."[2]

Corpus Christi entrepreneurs had gone even further, so determined to become the official anchorage for the Gulf that they persuaded the US Army head of engineers to override his own board. Arguing that the time consumed building a future "deep water port with proper railway feeders" would cripple present trade, General Marshal requisitioned an extra $100,000 for the corps to scour the route Kleberg had suggested—from the pass along Mustang Island to Corpus Christi. In the meantime, the city began designing a seawall and, with the full support of Congressman John Garner, strengthened its drive for an intracoastal canal.[3]

But in spite of trestles out of Rockport and channels through Turtle Cove, much of the uproar assailing Aransas Pass was caused by a new namesake, the Aransas Pass Channel and Dock Company. Inspired by its predecessor, the Aransas Pass Harbor and Improvement Company, but determined to avoid its errors, the San Antonio–led corporation chose to actively develop the pass. Even as Picton completed work on the south jetty, the company hired its own dredger to dig a trench twenty-five feet deep and one hundred feet wide from the town of Aransas Pass east across Aransas Bay. Propelling itself through reefs and sea-grass meadows, the steam-powered vessel shoveled into their bases, ground their matter into bits, then disgorged it sideways. Towering in ever-higher layers, the spewed soils settled into solidity, becoming the foundation for the company's next effort, a new railway connecting mainland wharves to Harbor Island.[4]

Excavation did not stop at island's edge, however. In May 1909, corporate spokesmen closed a deal with Powers Southern Dredging Company. The channel, already cut "in a direct line from Aransas Pass," would continue "between Mustang and St. Joseph's . . . to the jetties."[5] Harbor Island was to be split asunder.

As were shorebird nests, seabird aviaries, rodent holes, and snipe haunts along its center. Algae and worms clinging to edge-water flats disappeared, as did some snakes and amphibians. Decomposing shellfish reeked on dredger-dumped mud mounds. Construction continued unabated. The pass was open, an international harbor was coming, and once that ten-mile-long waterway cut through Harbor Island to city wharves, a spokesman was certain the government would "make . . . Aransas Pass at once a deep water port."[6]

No one else agreed, least of all communities surrounding the little town. Rockport particularly took umbrage, assuming the long-used channel from the pass that followed along St. Joseph's Island to its mainland wharves needed only extending to make it perfect for the army-built anchorage. The town "has a safe harbor," one spokesman noted, "and possesses fine terminal facilities both for dock and railroad purposes." Moreover, its leaders vaunted long-held ties with their sister city on the Guadalupe River. "The Victoria country . . . will, in the event that we secure deep water connection with the Pass, bear the same connection to Rockport as the Houston country does to Galveston."[7]

Corpus Christi went even further. Port backers were not only extolling deep water as far north as Corsicana but were also proselytizing commercial clubs and lobbying politicians for their cause. City mayor Roy Miller took particular pleasure in reporting Congressman John Garner's support. "It means that . . . Corpus Christi's future as a deep water port will be assured. . . . It is a great victory for Corpus Christi—the greatest in her history." It was the town's bluff, however, that provided its greatest advantage. Rising "forty feet high from water's edge," it overlooked "one of the safest harbors in the world," an editor enthused. A meteorologist concurred: "Corpus Christi is the safest place along the entire coast . . . topographically the site of Corpus Christi cannot be equaled."[8]

But securing deepwater designation required more than political support and scientific scrutiny—it required money, and the city of Aransas Pass had it. Now backed by a Baltimore syndicate as well as the Port and Harbor Company, its leaders broadcast their goals: "cutting a ship channel through Harbor Island to Aransas Pass . . . dredging a turning basin in front of the town . . . constructing . . . docks, warehouses, elevators, etc., along with railroad tracks," and then connecting those tracks to communities like Robstown and Laredo and San Antonio.[9] They offered speedy shipping to area businessmen frazzled by Galveston sluggishness. They promised "no-lightage offloading" to shopkeepers sick of paying twice to get goods ashore. They guaranteed cheaper fees to farmers crippled by differential freight rates. They coordinated with Congress, providing "inland improvements . . . and connecting railroads" while "the government continues in . . . completing the jetties." They published maps even more biased than that of Rockport and enticed newcomers with business lots selling for ten dollars down.[10] Most importantly, although they gained patrons from all parts of Texas, they maintained a special relationship with their San Antonio financiers. A Galveston reporter termed Aransas Pass town "an outpost of San Antonio and from its parent city has imbibed that spirit of progress that overcomes all obstacles."[11]

That such spirit could also be as fleeting as sea froth was rarely acknowledged, although District Fourteen congressman James Slayden voiced concern. Too aware of the risks of boomtown buying to stay silent, he warned his constituents of pipe dreams. "I am with you in the development of canal, river, and port, heart and soul," he assured them in late 1909. "But

Harbor Home

Aransas Pass's ad, extolling its availability as a "Gulf Coast Port." San Antonio Light and Gazette, May 22, 1910.

... don't rely upon ... commerce [with such as] the Panama Canal. ... Trade will not come without an effort."[12]

They ignored him. So convinced the army would designate Aransas Pass as the site for "the government port on the Intercoastal Waters of Aransas Bay," the people of that town went ahead with plans to build a schoolhouse, set up a commercial club, and publicize themselves. Stakes were driven into the bay for a new wharf and workers started preparing ground for railway trestles.[13] On October 20 they opened their Deep Water Celebration. "An army of home-seekers, investors, and strangers from all parts of the country has invaded Aransas Pass," a San Antonio paper reported, "[while] powerful dredgers have been cutting their way from the waterfront of Aransas Pass city to the open Gulf of Mexico between the jetties." Supremely certain of harbor affirmation, "today Aransas Pass is wearing festive attire."[14]

Rockport was unimpressed, confident that its eight-foot-deep channel from town to the pass would earn it the government's designation of port city, while Corpus Christi basked in its now deepened twelve-foot cut between the jetties and home. No one at all considered the little town still at the tip of Mustang Island, nor had it put itself into the running. Still smarting from Colonel Ropes's betrayal, the community had renamed itself Tarpon, after its most famous game fish. Practically atop the Aransas

Pass, the town had neither the means nor the inclination to fight to be a port. That battle lay with its neighbors, "the world outside."[15]

But when Lieutenant Colonel Lansing Beach, Major Charles Riche, Major Henry Jervey, Major James Indoe, and Major George Howell convened in Galveston in November 1910, after an intensive tour of the coastal barriers, their main concerns were distances and expenses: "$3,400,000 to construct a twenty-five foot channel to Corpus Christi, $2,000,000 to construct one to Rockport, and $600,000 to deepen the present Aransas Pass channel to twenty-feet."[16]

So the Board of Engineers did a King Solomon, ignoring the suppliants and slicing up the victim—no city would house the new deepwater port. Instead it would be carved into Harbor Island at "a short distance within the entrance to Aransas Pass." There, to correspond with the "depth already adopted for the channel in the pass between the jetties," the basin would be deepened to twenty feet. A channel would be built adjacent to it along with a roadstead into which boat slips could be cut, and a dike would be leveed on St. Joseph Island to prevent north jetty erosion. Since unencumbered right-of-way was a given, the entire cost of the government's work, not counting maintenance, amounted to a mere $375,000.[17]

Tarpon was exuberant: only eight hundred yards from the closest edge of Harbor Island and less than two miles from the proposed harbor, it could practically call itself a port city—and it did. Within months, having renamed itself after the pass, it would delight in its new glory: "Our Future Greatness Assured," "Port Aransas is really the Only Deep Water Port in South Texas," and "Port Aransas . . . got . . . away with the Lion's share of the appropriations." Rather than removing itself from its neighbors as in the past, the *Port Aransas Post* assured its readers, the community would now consider itself "in company with the live Gulf Coast towns that are doing things."[18]

But rather than doing things, Rockport, Aransas Pass, and Corpus Christi—those "live Gulf Coast towns"—were still in shock from the December 1910 announcement. Six months later the board struck again, this time accusing the three of attempting to monopolize rail lines to the newly designated Harbor Island Basin. It was not until the state of Texas, under federal pressure, guaranteed free harbor access to all regional users[19] that construction on the long-desired deepwater port finally began—making a significant impact on that tiny deltaic growth formed with the rise of Mustang and St. Joseph's Islands so many millennia ago. Composed of deposited sediment, Harbor Island rested between two natural waterways, its northward-flowing Lydia Ann Channel eventually becoming the classic route to Aransas Bay. Now, with its other outlet, Turtle Cove, deepened southward and its central innards sundered by the Aransas Pass Channel and Dock Company, only the islet's eastern edge appeared intact.[20] Still accreting sands, a gentle sturdiness there prevailed—even with the arrival of the *Comstock* and the *Guadalupe*.

The *Guadalupe* had originally anchored several thousands of feet east. Its task, to destroy any remnants of the old Mansfield Jetty, went methodically, each exploding shot of dynamite marked by twenty-foot-high eruptions of marine life and mud. It was the vessel's movement inland, however, that affected the isle itself. Joining forces with the government-deployed *Comstock*, it chunked out additional seafloor in preparation for an even

Dynamiting the remains of the Old Government / Mansfield Jetty. Courtesy of Port Aransas Museum.

US dredge boat *Comstock*, built for use in Aransas Pass waterways. Courtesy of Jim Moloney.

bigger dredger. Finally arriving, and commissioned at a significant price, that behemoth slowly sliced out of Harbor Island depths an anchorage eight football fields long and seven wide. Additional scouring—a turning basin, one channel lengthened north, and another one south—altered the eastern edge. It was the subsequent removal of a last bit of shoreline, however, to "extend the harbor front . . . at relatively small cost," that finalized the islet as a deepwater port—a status enhanced by the Aransas Pass Channel and Dock Company.[21]

That company, having earlier dredged the central channel, cut alongside it a minor one to access the dock, pier, and compress now lodged on its bank. Gargantuan and powered by steam, the compress crushed cotton bales into the hard-packed cubes necessary for ship loading. Grindingly noisy, it dominated the wharf, almost minimizing the rotund storage tanks lining the southeastern side and the pipelines linking them to boat slips.[22]

Possibly nothing impacted the isle as much, however, as the Aransas Harbor Terminal Railway. The system had been a vital part of the company's deepwater plan even before 1911, and although the army's ultimate choice of Harbor Island rather than Aransas Pass town had shaken investors, the company quickly regrouped. Dutifully accepting state and federal directives prohibiting monopolies, it proclaimed its welcome "for any railroad to enter the field,"[23] then set up a single train line combined with

Harbor Home

a drawbridge to the island wharf. Employing the piled sediment earlier dumped by dredgers, company engineers strengthened the mound into a two-hundred-foot-wide embankment alongside the channel. Upon this and through it they drove trestles, drilling out into water when they edged off the flats. For over six miles the new rails spiked bay floor and spoil banks until finally merging with the San Antonio and Aransas Pass terminal on the mainland. From there, in little over a year, sixty-seven thousand bales of cotton chugged eastward to the new harbor basin. There, offloaded, pressed, cubed, and refreighted, some were shipped to the northern Gulf, others to Europe.[24]

Petroleum also moved, over one million barrels imported from the Mexican coast to island storage units in tankers. Spawned in the late 1880s to carry oil from drilling fields to refineries, these tankers had evolved internationally from huge, seven-masted schooners to steam-powered storage vessels holding up to twelve thousand tons of crude oil.[25] None that vast powered through the pass, but their smaller versions crossed, captains and pilots alike relishing its corps-increased width and depth.[26]

Nudging the carriers through the harbor into their berths were other craft new to Aransas waters: tugboats. Deep hulled, with rounded bows overhung with mats, the single-engine vessels could be less than seventy feet long and yet combine to move tankers and barges six times their size.

Tugboats—lithe, maneuverable, and vital to in-harbor traffic. Courtesy of Port Aransas Museum.

Already central to upcoast harbors—pilot Tom Mercer had been mastering his own tugs in Galveston for years—tugboats now became basic to Harbor Island.[27] With their help, ships could maneuver within the basin with as little encumbrance as they had when they entered and exited the pass. Soon an army engineer's fear that the new port's commerce had not yet "attained the magnitude expected for it" proved unfounded. Besides the traffic provided by at least three oil companies,[28] a coastwise shipping line started coming from Galveston, a consortium of foreign steamers provided stopovers for cotton, and Seaboard and Gulf Steamship transports began arriving at least twice every month. A six-barge consortium opened on-site, substantial enough to haul seven hundred tons of freight per vessel to area towns and cities. Within two years of the basin's opening, oil tanks dotted Harbor Island's coastline, boat slips slotted its edge, and switch systems, boxcars, and warehouses lined its artificial channel.[29] Those original inhabitants that could—waterfowl, marine life, and land animals—moved elsewhere.

CHAPTER 14

Hunters and Hard Hulls

Many moved to Mustang Island. Its dunes were livable, and its shrub areas gave ibis and egrets room to lay the eggs necessary for preservation. The island's sand expanses, spread so widely lengthwise they ensured privacy and isolation, promised haven to black skimmers and terns. Ducks thrived, redheads feasting on shoalgrasses in ponds as readily as scaup and pintails.[1] But even as they found sustenance away from ravaged tidal flats and marshes, other threats reigned, for Mustang also attracted hunters.

Coastal ponds, estuaries, and lagoons were natural refuges for fowl heading south for the winter; when they swooped down to drink, shooters

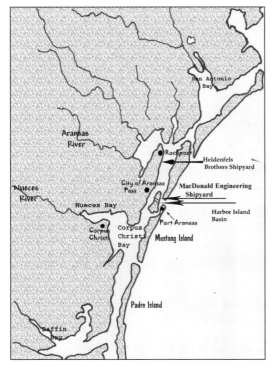

Major Coastal Bend communities, 1910–1919.

Hunters with their bounty. Courtesy of Corpus Christi Public Libraries.

opened up from below. Hunters in the 1890s rejoiced in their bounty as thoroughly as had their Mexican War counterparts, and Galveston papers commented on "Aransas Pass waters . . . alive with ducks." The area resounded with "gun shots," a reporter noted, "heard at all hours, in town and out on the bays." Even marketers took advantage of the abundance, slaughtering an average of thirty thousand canvasback ducks a year and selling them in San Antonio.[2]

But it was to another kind of sportsman that the island appealed as well. For with every waterfowl enthusiast willing to buy ammunition, hire a guide, and rent a boat, there were many more willing to fish—and despite the intrusions of jetty building, seafloor dynamiting, and harbor dredging, Port Aransas had fish. Mullet were so plentiful they could be harvested locally. And menhaden were also abundant enough to be commercially profitable. Redfish and drum, scavenging among the shallows for crustaceans, made themselves easy targets for enthusiasts, as did sea trout and catfish in nearby shoalgrass meadows.[3] It was the larger fish, however, spawning inside island estuaries, thriving within lagoon waters, and emerging majestically mature in the pass and in the Gulf that attracted

Hunters and Hard Hulls

big-time fishermen. One of the most memorable was the goliath grouper, whose tendency to cluster "among the brown sea-mossed rocks of the Aransas Pass jetties" was matched only by its large size. A three-hundred pounder, caught by a friend of Colonel Leonard Wood, was dressed, iced, and shipped to San Antonio, where it provided "one meal for the entire [Rough Riders] regiment." Others were caught, secured in pooled areas within the bay, and, rather like the green turtles penned up decades earlier, "kept alive until needed . . . then hoisted out by a block and tackle and butchered like a hog."[4]

The tarpon, though, earned Port Aransas the title "Paradise of Sportsmen." These giant silvery herrings, spawned in shallow waters off the pass and nurtured with estuary shrimp, crabs, and mullets, became one of the most renowned fish of the western Gulf, often measuring as much as eight feet in length and weighing hundreds of pounds. It was their fight, however, that made them notable. Once one was caught on a hook-embedded mullet, it battled to dislodge the bait, twisting, flailing, wrenching its jaw crossways while propelling itself upward and out. Flying "above the surface with a force that sets the foam curling . . . mounting the waves," trying again and again to "shake loose the steel prong," its "magnificent proportions break the water," one witness exclaimed, "leaping eight and ten feet into the air."

A goliath grouper displayed atop a Rockport pier. Courtesy of Corpus Christi Public Libraries.

The "silver king" leaping against a backdrop of jetty and birds. Courtesy of Corpus Christi Public Libraries.

Striving for release, gasping almost lunglike in its airborne jolts,[5] diving torpedolike into the depths, then rising to streak out again seaward, the tarpon reigned as gladiator of the Gulf, a challenge for the most avid fishermen.

And they met it, flocking to shallows in the Laguna Madre and to rocks near Aransas Pass with boats, reels, rods, and forty-pound lines. For many, catching tarpon was like counting coup, seeing how many one could land in a given period. "How would you like to go out and tear the record for light tackle tarpon to pieces?" a cry would echo around Port Aransas piers, and fisherfolk responded. One, J. S. Ingram of Pine Bluff, Arkansas, brought in seventy-six "silver kings" in forty-six consecutive days. "He also brought in nine successive tarpon upon nine strikes, not missing a fish," an admiring fan added. Others, however, simply desired the intensity of combat: "to feel the mighty thing tug at the slender line, to hear the whir of the revolving reel, the sudden swish of the rod . . . [to see the eruption] from the deep green water of a magnificent vision in green and silver . . . [to experience] the cruel steel of the gaff entering its brain." One statement captured it all: "To witness the splendid breaking of a wounded tarpon is in itself . . . glorious."[6]

Guiding sportsmen to their quarry became a career for turn-of-the-century islanders, but a career vulnerable to the waves and winds of offshore fishing fields. Choppy, twisty, and variable, they made trolling difficult in the cumbersome rowboats originally used. Early motorboats like that employed by St. Joseph's Ned Green to convey Tarpon Club friends to favorite spots were quicker but just as unwieldy when caught in rough seas. It fell to Mustang newcomer Fred Farley to design the perfect island vessel: a low-sided eighteen-footer with a chine (that point on the hull where sides and bottom intersect) high off the water. Powered, as likely as not, by a Model T engine, stable enough to crush fractious waves, but streamlined enough to slip through their troughs, the vessel was a boon to fishermen and hunters alike. In time Farley and Sons Boat Works became as intrinsic to the little Mustang Island community as its fishermen, towmen, pilots, and bait dealers had become. The coming of the port only enhanced the citizenry. By 1912 Port Aransas had doubled its number of certified voters; added butchers, on-site mechanics, and jewelers to its commercial club; and incorporated itself into a city commission form of government. Numbering over two hundred residents, anticipating its own corps-approved subharbor, and adjusting to a "floating saloon" outside its limits, the town was booming.⁷

As was the city of Aransas Pass, buoyed by an enviable linkage to major railroads. Its San Antonio and Aransas Pass line, with those eastern tracks across the bay to Harbor Island Basin, also had connections to the Sunset Central system, the Missouri, Kansas, and Texas line, the International and Great Northern Railroad, the Frisco system, the Santa Fe line, and

A Farley boat, at home on the water. Courtesy of Port Aransas Museum.

the Cotton Belt railway. This put it "more in touch with all Texas posts than is common for cities having only one railroad," an advocate noted. Its unique access to the port also brought it a busy commercial wharf, another cotton press, several lumber companies, two banks—and solid ties to San Antonio investors.[8]

Native wildlife contributed as well to the town's prosperity. "The prairies sloping back from the bay furnish exactly the sort of food for geese and brant," a writer noted, referring particularly to the grasses and wild onions available on the mainland. Hunting remained primary, as it had on Mustang Island: "Aransas Pass is peculiarly situated in the matter of fresh water lakes, there being two or three good-sized bodies of water within easy riding distance from town, where the best of shooting may always be had." Commercial fishing flourished with a canning factory planned in 1911, and shrimp continued to be harvested during the summer. Redfish Bay, on which the town fronted, promised the sports angler "a great variety of edible fish."[9]

But it was the innate optimism of the city, nurtured through seemingly innumerable town lot sales and rallies, that characterized it most. "Aransas Pass has many natural advantages," a chamber of commerce leaflet read, "among which are a delightful climate, an abundance of good water . . . natural shade trees, and good drainage. . . . We have . . . two miles of concrete sidewalk . . . several miles of improved streets, a twenty thousand dollar public school building . . . [and] six churches." More important than anything else, however, were the town's investors, not just outsiders but local men and women—candy shop operators, grocery store managers, general merchandise proprietors, bakers, stevedores, boatmen. They all owned a stake in their community and its connection to the port. It was they who brought its population close to two thousand by 1912, an accomplishment "truly wonderful," even if Aransas Pass leaders said so themselves.[10]

But not all Coastal Bend municipalities appreciated Harbor Island Basin; Corpus Christi was the most disgruntled. Situated on the edge of the same bay that carried Taylor's steamers and Kittredge's ships decades earlier, the town had grown nearly twenty times over its 1846 population. Its bluff, on which Colonel Kinney had posted cannon, now housed the elite of South Texas, widow King and the family of Mifflin Kenedy among those owning mansions overlooking the town. Shoreline businesses crowded the esplanade on which Hitchcock and Whiting had drilled their men, and a new county courthouse stood several blocks north of the spot where townspeople had lynched Northern sympathizers. A modern sewage system utilized the same Nueces River from which soldiers had carted fresh water, and oil agents and wildcatters scouted the plains where young Grant had ridden earlier.[11]

Sportsmen angled for tarpon within Flour Bluff's waters almost as avidly as they did off the jetties, and the nascent Corpus Christi Fish and Packing Company was expected to rival any competitor on the islands. The military had sited a rest-and-relaxation camp along the North Beach peninsula for those guardsmen still protecting the border, and the largest hospital in that part of South Texas sat on the same beach, established by the Sisters of Charity in 1915. A new causeway crossed Nueces Bay where the old oyster reef had lain, and arrangements were being made to try to deepen the city wharf.[12]

Harbor Island Basin was still valued, however. Locally based fisherfolk and dredger captains had to have noticed an upturn in business once it opened, and the deepened channel through Turtle Cove did bring in more trade, although the shallow bay floor remained a problem. But with a population approaching nineteen thousand by the midteens, an economy sustaining two major banks, and an agricultural base supporting four separate railroad lines, "the matter of getting deep water for the city [was becoming] a matter of . . . tremendous importance." Corpus Christi, its supporters agreed, deserved to be the area's premier deepwater port.[13]

So despite warnings to avoid "pulling wires in Washington," city leaders resorted to every stratagem available to supplant Harbor Island Basin. Roy Miller was especially adept, his years as mayor, as advocate for an intercoastal canal, and as Washington insider giving him heightened political skills. Thus, a 1915 bill giving the city total control of the "land lying under the waters of Corpus Christi bay and within the limits of the City" became a four-year focus of his energies, going through several variations (one a version sponsored by Padre Island's Pat Dunn, who had entered politics by this time) and incurring a rival politician's wrath before it came close to agreement in early 1919.[14] Miller's reason for so relentless a drive: the War Department was intending to "deepen [Corpus Christi's] channel and harbor," provided the city submitted a "comprehensive and adequate plan for [such] municipal terminal facilities." He announced the news just as the bayfront issue approached resolution, revealing not just his personal ties with government officials, who supplied him advance information on the army's plans, but also the necessity of securing immediate city ownership so that deep-harbor development could start. Buckling under pressure both local and national, the opposition gave in, and on March 29, 1919, civic leaders could state: "DEEP WATER . . . IS ON THE WAY; IT IS ALMOST HERE."[15]

Rockport residents were unmoved, especially those who kept up with local politics like Charles Johnson and Harry Traylor, both former mayors. By late spring of that year, everyone knew that Roy Miller would lose the next election; with him would sink Corpus Christi's hope of deep water.[16] But that mattered little to the people on Aransas Bay's outcropped shore. Though they had lost the port, they had gained better channel access, a sure passage to the jetties, and elevated status as "Uncle Sam's new resort on the Gulf." Consequently, townspeople organized en masse to defeat a state railroad commission attempt to destroy Harbor Island Basin. "Powerful interests have fought this harbor and attempted to block its improvement for more than twenty years," the head of Rockport's commercial club asserted in 1912. "Now that its location has been definitely and forever settled . . . the next question is, how best to . . . utilize it."[17]

Utilize the deepwater port Rockport did, once railroad opposition had been thwarted. The shrimping industry regained strength. Watermen congregated in the town, tugmen, pilots, stevedores, and longshoremen beginning careers in its bays that would last a lifetime. Musicians crowded Bailey's Pavilion in the summer months, attracting partygoers from all over the Coastal Bend. Tourists flocked to the beach as they always had, now able to attend open-air theaters, road shows, and saloons. Even the Great War, in spite of the death and disease it caused,[18] brought dreams and hope to the town.

As it did to Harbor Island, at least in its aftereffects. World War I itself hurt the port, in trade most especially. Early on, commerce had been so steady that the government set the basin's freight rates at parity with those of Galveston; annual traffic by the end of 1914 totaled over 192,000 tons.

But conflict overseas stalled shipping to the Gulf, and by the end of 1915, the port's activity had dropped over 10 percent. The next year saw an even bigger decrease in trade, fuel oil from Mexico being the only imported item noted. In 1918 there was a slight increase, much of it crude oil again sent from Mexico; that product was so cheap, however, that it dropped the overall value of port traffic to 60 percent of its earlier output.[19]

But it raised the value of Harbor Island—or at least of its shoreline alongside the channel leading to the pass. That locale made a perfect site for a new kind of construction: concrete ships. Troubled by the corrosive effects of crude oil on the inner linings of tankers, designers had been searching throughout the war for any material that would protect the holds and prolong the use of these necessary vessels. By 1919, the MacDonald Engineering Company seemed to have succeeded, adapting an old French practice of coating metal mesh with a plaster-like paste, a technique called ferro-concrete. Creating two vertical intersecting cylinders—framed in steel mesh, layered with portland cement, and allowed to set—engineers

The concrete ship *Darlington*, hitting the waves at its christening. Courtesy of Texas Maritime Museum.

then turned them horizontally and slipped them between an already-assembled bow and stern. Once seven double-cylinder sections had thus been aligned, slabs were laid, one above for the deck and one under for the keel. The deck slab featured a large bridge and a funnel extending down below, belching out smoke from the engine room's two semidiesel engines. The keel sported, in addition, oil ducts that pumped the crude out at port and a boat-long passageway where the cylinders intersected. Almost as important as their construction, however, was the size of the ships built on the island. Less than three hundred feet long, neither the *Durham* nor the *Darlington* took up more than eighteen feet in depth when it crossed Aransas Pass.[20]

Almost taken for granted as the tankers took shape were their navigational tools. Sextants were still used in determining latitude at sea, albeit far more streamlined than those used by Pineda or Bérenger. Chronometers, developed in the late 1700s to accurately measure distances from east to west, were also more compact. But gyroscopic compasses had replaced their magnetic forebears, guaranteeing true north as opposed to that affected by magnetic fields, and stadimeters had become helpful in determining the location of other ships.[21] It was the radio transmitter, however, that generated the most notice. Developed in the late 1800s to convey the same speed of communication to ships at sea that the telegraph provided on land, wireless telegraphy had made flag signaling almost obsolete. Utilizing atmospheric radio waves, Guglielmo Marconi had learned how to capture and send them in staccato-like pulses. With specially designed equipment, an operator innumerable miles away could retrieve them from the air and convert them into messages: ships sailing close by, icebergs looming ahead, winds churning the sea. By the time MacDonald Engineering started constructing its first boats, industrial countries had mandated international code agreements and the United States had required all large ships entering its ports to have radios. As the *Durham* passed through the jetties into the Gulf, it dutifully sported an antenna on its mast.[22]

Alas, the *Durham* and the *Darlington* faded quickly into shipping history, concreted hulls not as effective as hoped. Less quickly shelved were the dreams of Rockport residents, who had invested in their own boat-building industry. These hopes had formed in early 1917, when the brothers Heldenfels, lumberyard owners near the Aransas River, petitioned the community to help them construct seagoing vessels for the war. Eager for the business, town fathers donated thirteen acres near the end of Water Street for the yard, $30,000 to defray charges for equipment, and significant political pressure on the US Shipping Board for government support. Intrigued, its Emergency Fleet Corporation delegated an official to the Coastal Bend, who promptly fell in love with the place. Rockport provided, he proclaimed in a local interview, an "ideal" location: fronting a bay "only ten feet deep . . . immune from submarines."[23]

So the brothers started building and skilled laborers came for jobs. "At the peak of construction," one source reported, "over nine hundred men worked at Heldenfels Shipyard." Some were locals, eating their lunches under "a big roof . . . made of palm trees . . . for shade." Others were more transient, happy to hike to rooming houses for meals. But their presence cheered the town, even when the Emergency Fleet Corporation reneged on its original order. Despite disappointment, the men gathered on the last day of July 1919, along with Governor William P. Hobby, his cabinet, and a whole host of dignitaries, to dedicate the brothers' first ship, the *Baychester*. The community stayed cheerful. In just five more weeks, its second ship, the *Zuniga*, would be launched, and after that, at least two barges were in the offing. In spite of federal recalcitrance and a European armistice, the future was promising.[24] Although always vulnerable to the winds of chance, Rockport believed good times lay ahead.

CHAPTER 15

Hurricanes

Winds—of nature—had always driven the barrier islands. It had been wind from the northeast, in months like September and October, that had pushed longshore drifts southward, bringing sediment from up-country rivers to the base of the isles. It had been wind during March and May that washed Rio Grande sand northward onto island deltas and mudflats. It had been unrelenting wind that formed dunes on the emerged barriers; it was wind that forced runnels between them and their hummocky mounds. It was wind that spurred the thunderstorms driving Narváez and his ships off course in 1528 and nearly capsized General Dana's fleet three hundred years later. Wind created northers, that Coastal Bend phenomenon that "swoop[ed] down," held "high carnival" among

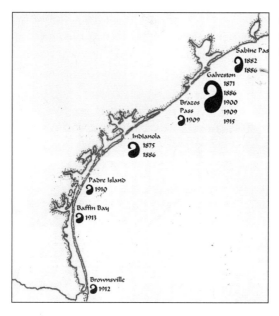

Coastal areas damaged by hurricanes, 1871–1915. Based roughly on David Roth, "Texas Hurricane History."

Hurricanes

bay waters bursting over banks, then abruptly changed direction to strike from the north. One sank La Salle's *La Belle*, another struck Dana's men as they attacked Fort Esperanza, and a later norther ruined Aransas Pass's Deep Water Celebration.[1]

But most of all, on the barrier isles of the Coastal Bend, wind meant hurricanes, fierce cyclonic sea-spawned maelstroms, so violent that they tore across oceans and spiraled onto shorelines, wracking and inundating all in the way. Formed by a combination of low-pressure turbulence, hot rising air, and upper atmospheric instability common along the tropics during certain months, they haunted Robert Mercer, worrying about victims after the 1871 hurricane. They destroyed the town of Indianola off Matagorda Bay, hitting it twice in little over a decade. They surged through Brazos Santiago in 1844, slashing a ragged hole right across the tiny settlement.[2]

Hurricanes demolished nests, destroyed aviaries, swept through seagrass meadows, and ravaged estuaries. They poured seawaters over coastlines and cities, flushing out fresh water and salting rivers. They coursed over pastures and drowned cattle and horses. They flooded beaches, demolished mudflats, and pushed fish onshore too long to survive without oxygen. Hurricanes cut swaths through landforms and ripped new entries into bays. They recast nature.

And, as part of nature, they altered Coastal Bend life. Some inhabitants, like the townspeople of Indianola, seemingly dared hurricanes to hit, first moving off high ground to sea-level sands in 1854, then remaining beachside after the 1875 devastation. Ultimately abandoned by inland financiers, they refurbished ruins with promenades and resorts, creating a "Dream City on the Gulf," until a last blast killed that too. Its demise, however, created opportunity for others. Corpus Christi promoters in particular quickly linked the catastrophe to an economic rival: "It was an appalling disaster and . . . the conclusion is reasonable that some time or other, a tidal wave, driven by some howling tempest as swift and terrible as what swept over Indianola, may yet engulf Galveston Island."[3]

Galveston supporters flared back, deeming such comments demonstrations "of hostility toward a city which has always endeavored to cultivate the good will of others." But even then the Queen City created its own association with disaster: "There are those that would rejoice to see Galveston with all her inhabitants meet the fate of Sabine Pass," a site hit twice since 1882.[4] But the metropolis, with its beachfront bordering the Gulf on the east and its wharves extending down into the bay on the west, had an undeniable record of hurricane hits.[5] Although only the most recent—those of the 1870s and the 1880s—tended to be remembered, Galveston needed its own guarantors of security if it was to keep its rail lines and shipping trade busy and its port full.

Euphemisms helped; tropical storms were "quite a blow," and inundations were simply "overflows." But most persuasive and definitely most appreciated were statements by oceanographers like Matthew F. Maury, who maintained that "Galveston was exempt from the force of destructive hurricanes because it was located in a 'cove of safety,' protected by shallow water and sandbars running parallel to the shore." Others agreed, with possibly the most authoritative analysis written by Galveston's own meteorologist Isaac Cline. Thoroughly convinced that the two hurricanes that hit Indianola were "accidental" and that his city, "on account of its location and the peculiar features of the gulf coast, is not liable to be caught in the track of any . . . hurricane," Cline emphasized one more deterrent: its bay. "As there is too large a territory to the north which is lower than the island, over which the water may spread, it would be impossible for any cyclone to create a storm which would materially damage the city." The water in all its vastness would absorb any shock from the sea.[6]

So it was with no little concern that the *New York Times* reported, on September 10, 1900, "that an appalling disaster has befallen the city of Galveston." Follow-up dispatches estimated "two thousand lives . . . blotted out . . . tremendous property damage incurred . . . and the lower portion of the city . . . all under water." More articles appeared, and within days the nation read of the Orphan Asylum, which had fallen "like a house of cards," and the patients in St. Mary's Infirmary, "only eight of which . . . have been saved." Descriptions sickened: dead children tied to horses, women drowned while giving birth, bodies beaten by debris and stripped of all clothing—horrors increased with each edition. And as the public cringed, civic leaders all along the coast struggled with one unavoidable fact: rather than protect the island, as Cline had predicted, its bay had betrayed it, flooding its west side with wind-swollen waters even as surge tides swept over the east. Caught between the two, Galveston crumbled.[7]

The Coastal Bend was not unaffected. A relief boat from Corpus Christi

Galveston survivors cleaning up after the 1900 hurricane. Courtesy of Texas Maritime Museum.

was among the first to reach the city after the hit, and dances were held later, the proceeds to benefit survivors. Rockport sent supplies, too, and worried about citizens who had been visiting the island. Mustang's shoreline reeked of piled-up victims: "The entire beach," one paper reported, "is strewn with dead animals—cattle, horses, sheep, goats, etc.," while Captain Tom Mercer related his own travail at Galveston's port. After unloading a cargo that forbiddingly blustery morning, he headed for shelter in the Cotton Exchange building. There, "I remained," he remembered, while "the wind and the noise from the roofs rolling up kept up a continual roaring nearly all night."[8]

But more vivid than the memories and images was the lesson learned: a city on lowlands facing the Gulf was doomed. Port Aransas was a quick study, touting its location "on the north end of Mustang Island, the highest island on the coast," and admonishing the "knockers on the mainland [who] tell you it is unsafe to live on the island because it overflows, it is an utter falsehood! . . . Mustang Island has about 20,000 acres of land that lay . . . from one to thirty feet above high tide." Even disgruntled neighbors had to admit, "No storm has ever swept this little town off the face of the earth."[9]

Rockport was even more explicit with its "landlocked [and] unassailable . . . safe Harbor," and Aransas Pass, while extolling Harbor Island Basin's security as "the best and safest on the Texas coast," also touted its own. "The town . . . has one of the finest locations on the Texas coast for a great city. There are five great tiers of islands between its location and the Gulf of Mexico, giving it perfect protection from all Gulf storms and ocean waves."[10]

But as the most comprehensively sheltered community on the coast, Corpus Christi claimed the title. Its location provided one safeguard, a writer noted: the city was "situated outside the hurricane belt. The big blows originate in the West Indies and sweep northwestward toward Galveston, leaving Corpus Christi in a cup-like eddy off the left." Even the coastal plains to the west were an advantage. "Another point in favor of Corpus Christi," a meteorologist added, is that "there is no water in back of the city . . . nothing but good, hard solid prairie land." There would be no Galveston Bay uprising to sweep crossways across this Coastal Bend city. Little Nueces Bay to the north could even prove a defense. In case of a terrible storm, the bay's "great volume of water . . . would draw Corpus Christi Bay" toward it, not the town. But this was idle speculation: "The chance of such a storm striking this city is, to say the least, mere chance."[11]

Even more protective was the black clay headland that towered over the bay, the "bluffy banks," one called it, "that rise to a height of forty-five to fifty feet" and are "the highest . . . on the entire coast of Texas." Any waters that flooded the beach could surge no farther inland than a few hundred yards, "for there the bluff begins—rising almost precipitously . . . thirty feet or more, and there life would be safe from drowning." It was this bluff, "high above the reach of the tidal storms," that made Corpus Christi "the safest for our wives and children to visit during the summer months."[12]

The most persuasive argument for Corpus Christi's invulnerability, however, was the most ironic, for, as those in Aransas Pass had done, supporters cited the barrier isles as their last great defense against a storm. It mattered little how island residents touted their own immunity; to mainlanders, they seemed to exist mainly as external shields. Observers like W. L. Coleman of Huntsville, Texas, believed it: "That line of islands separating Corpus Christi Bay from the Gulf protects from all such cyclonic tornadoes as almost destroyed Galveston." The editor of the *Caller Times* agreed: "Such a disaster could not possibly befall Corpus Christi . . . being protected from the Gulf by Mustang Island, some twenty miles long and about two miles wide." Meteorologist George Reeder was even more explicit: "If such a storm should even approach this immediate coast, Mustang Island, a 'waif of the sea,' is directly in front of Corpus Christi. . . . Nature's Wall, placed there to receive and break the force of the mighty water that runs ahead of the storm. . . . Padre Island . . . separated from Mustang by a narrow pass only . . . is but a continuation [of this Wall]."[13]

Ultimately, whichever defense the Coastal Bend communities claimed—islands, elevation, or "cup-like" eddies—each rejoiced that "Nature has been lavish with her gifts to this section."[14] Man need do no more.

Rising from its own ruin, Galveston disagreed, as did Isaac Cline, writing two weeks after the devastation of his city. "I believe that a sea wall, which would have broken the swells, would have saved much loss of both life and property." Politicians, engineers, and citizens concurred, and within four years of the initial disaster, a seventeen-foot-high concave wall, laid over fifty-foot pilings and a fifteen-foot concrete base, faced the sea. In addition, five hundred blocks of the city—homes, churches, gas lines, and water mains—were elevated, with fill material scooped from the Galveston Bay floor and pumped in by dredges. The barrier was an engineering feat of immense cost, more than justified five years later when another hurricane swept through the area. "The great seawall has completely vindicated its efficiency," Galveston's mayor asserted. But it was the laconic observation by Corpus Christi's captain Will Anderson that proved the bulwark's worth: "If it hadn't been for the seawall, Galveston would have gone to destruction."[15]

Some jetties nearly did, like those at the mouth of the Brazos River, which lost several multi-ton rocks. That storm did damage as well to Aransas Pass's south jetty, carving a small inlet on its base. The next month saw another hurricane, and in autumn 1910, one with winds up to 120 miles per hour hit Padre Island, totally submerging it at landfall. Two years later, within a day after the *Caller* reported a hurricane too far away to fear, Brownsville was deluged, bridges at Point Isabel destroyed, the steamship *Nicaragua* run aground on Padre Island, and parts of the Aransas Pass Terminal Railway smashed.[16]

Disasters mounted as hurricanes continued to strike the western Gulf, three since the two in 1909. Still residents were unprepared when word came of a disturbance off the coast of Florida in mid-August 1915. A day later the storm hit Galveston with winds over one hundred miles per hour, surges twelve to fifteen feet high, and waves so rough they threw a four-masted schooner over the seawall. People drowned up and down unprotected shorelines as the hurricane tore inland, destroying as it went. "Over a zone reaching over a hundred miles on each side of Houston the vast flat of South Texas prairies is dotted with crushed buildings," one article read. Others guessed at death estimates: "Three Known Dead at Port Arthur," "Known Deaths at Galveston and Other Coast Ports Total 116," "110 Known Dead Result of Texas Gulf Storm."[17]

Yet, although townspeople on the Coastal Bend acknowledged that "weather conditions in a large portion of Texas were unusual Tuesday," the date was noteworthy locally only for its heat. In addition, the *Corpus Christi Caller* reported, its bay was "perfectly calm during the morning and afternoon." Although there was more trestle damage on the Aransas Pass Terminal Railway and some boathouses were lost off St. Joseph Island, there seemed to be few other effects. Already the south jetty on Mustang Island, a victim of the 1909 storms, had been repaired. A subheadline, appearing a day after the August 16 hurricane, indicated "No Damage Reported in the Corpus Christi Section."[18] Nature seemed to have demonstrated again her regard for the Coastal Bend.

Almost in confirmation, the next year brought glorious weather, such "beauteous rains" falling in July that the paper could headline another "Season of Plenty and Prosperity for the Entire Gulf Coast Country." Others around the state agreed, for August, the traditional month for family vacations before school opened, saw an increased number of tourists in Corpus Christi. Many had been in Galveston the previous year only to endure the 1915 hurricane; assured of Coastal Bend invulnerability, they relaxed in their beach camps and rooming houses along the bay.[19]

Until Friday, August 18. Late the night before, winds, typically southeasterly, had started coming from the north; by morning they were over twenty miles per hour. Rains began, at first light, then increasing with thundering rapidity. By one o'clock, authorities began warning roomers and campers along the bayfront to come inland, to take refuge in one of the city's bricked banks or a church.[20]

Townspeople up the coast reacted more swiftly. One, Aransas Pass shop owner Frank Clendening, had been reluctant at first. "We . . . had plenty of warning but never once thought of [the hurricane] coming here," he acknowledged. "We had been made to believe this area was storm proof."

Winds over eighty miles an hour by noontime changed his mind, and having secured his home, he struggled to his store on Commercial Street and "just stood at the front door and watched." Between fiercely driven rain sheets, he heard of Harbor Island Basin blown to bits, tugboats tossed against the railway, and terminal trestles going down. Sounds of ripping awnings, cratering roofs, breaking windows, and fragmenting storefronts shattered the night, and then the bay surged, breaking over "the Southern Pacific tracks and . . . coming up toward the town fast." Finally, at early dawn, winds and rain let up, revealing "the worst torn up town you ever saw," Clendening wrote his wife. "The San Antonio and Aransas Pass tracks are a mass of debris . . . the Port A mail boat is sitting right side up by Mr. Hundley's office . . . the wharf is a total wreck . . . and the terminal was washed completely away." Boards and mud clogged every street, and of the myriad light craft harbored in adjacent waterways, fewer than a dozen boats were left: "All the rest went down and were torn to pieces." Nine died in the immediate area, including several crew members of the steamboat *Pilot Boy*.[21]

The entire region suffered from hurricane rage, people as far inland as Jim Wells and Duval Counties dying in its aftermath. But it damaged the Coastal Bend most, with Aransas Pass's business section ravaged, Rockport's waterfront devastated, and Port Aransas reportedly "washed off the map." Its impact on Corpus Christi, however, was more complicated. Physically, the town was a shambles: its two resort piers were gone, its most elite department store had been unroofed, and its two-storied rooming houses lay in scraps along the beach. Moreover, all those tourists who had come, confident that they were summering in a sheltered stormproof haven, were now sleepless and clothes-less, going home with only their lives to be grateful for.[22]

For all its assurances that nature had blessed it, how was the city to deal with this? Even its neighbors wondered. "The report is that Corpus Christi was worse hit than we were," one wrote. "The causeway was washed out and the town flooded. Nearly every house in town was damaged. . . . But the papers just tell what they want. I will bet that Corpus Christi will not let the extent of the damage get in print." It didn't. The very day after the hurricane, city headlines read, "Corpus Christi Defies Tropical Hurricane." Subheadings told the tale: "Flimsy Waterfront Structures Crumble but Main Business and Resident Sections Stand Storm," "Not a Life Lost

Port Aransas's wrecked shoreline after the 1916 hurricane. Courtesy of Port Aransas Museum.

Shattered rail line and wharfs in Corpus Christi after the 1916 hurricane. Courtesy of Corpus Christi Public Libraries.

and Not a Serious Personal Injury of Any Sort Reported," "In Spite of Losses, Citizens Feel Happy."[23]

Corpus Christi had turned the disaster into a shaded blessing. Its people, "not at all discouraged at the fact that Corpus Christi for the first time in history, has had a severe storm, were pleased . . . that not a single loss of life occurred." Instead, they deplored the shoddiness of the structures destroyed and blamed the buildings' poor location. Little was noted of the concrete bridge over Nueces Bay now shattered or the warehouses on the waterfront in splinters or the display tower on Municipal Wharf beaten into the bay. Instead, emphasis focused on outlying communities, the losses they incurred, and the property damage they suffered. To Corpus Christi leaders, the 1916 hurricane was but one more "demonstration that [the] city is practically storm proof." It confirmed "the safety of Corpus Christi as a port of entry," a businessman expounded a week after the hit. "The time is ripe for continuous . . . effort" to secure an anchorage right here in this bay, he decreed, to replace Harbor Isle's vulnerable deepwater basin.[24]

That it was so vulnerable apparently escaped the people who lived on Harbor Island and its fellow isles. Pat Dunn, whose ranch house had been swept away by surge waters, was already rebuilding on Padre Island. Although Port Aransas fish dealer W. B. Harmon bemoaned the "big storm of the 18th" and the loss of his boats, he still had his "life and a few dollars in the bank," he wrote a supplier. Deep-sea diver Nick Kahl seemed to feel just as sanguine. Just a month after the 1916 hurricane he was pitching a project to colleague Tom Mercer, angling to raise bank vaults from Matagorda Bay. Mercer himself was ordering supplies and lumber within weeks of the strike on Port Aransas, confident of their timely delivery.[25]

He had reason to be confident: the San Antonio and Aransas Pass Railway had already repaired its tracks by August 31. Those tracks were not only carrying the few shipments of oil still making their way from the port during the war; they were also bearing a converted Model T truck designed to run on rails, thus providing a much-appreciated passenger service for workers commuting to harbor jobs.[26]

Nor did a sense of vulnerability deter the Heldenfels brothers from starting their shipyard in Rockport twelve months after the hurricane—or stop MacDonald Engineering from producing their concrete vessels on Harbor Island two years later—or keep the military from budgeting Harbor Basin expenditures for the oncoming fiscal year. The Corps of Engineers had already sent the *Sam Houston* there in the spring of 1919 to redredge "the deep-water harbor in front of the wharves" and was in the midst of formulating plans to "maintain the Harbor Island Basin and the channel to the town of Port Aransas" well into 1920.[27] Any chance that Corpus Christi leaders had of portraying the barriers as hurricane doomed was stymied by the islands' own energetic renewal.

So as the pungent stink of bloated bodies and burned carcasses slowly dissolved and the more usual smells of salt-laden air and sand returned, the Coastal Bend healed itself. Padre Isle took renewed nourishment from fresh sediment thrust upon it by the hurricane's surge; estuary and bay life burrowed new homes into sunken ship spans and driftwood strewn along Laguna Madre; and lightermen and Farley boat captains took note of sudden seafloor hollows and holes carved out by raging wind-cursed seas. Locomotives trundled along the mainland, bringing Great War veterans home; daredevil pilots entertained locals, taking off and landing on the flats next to Aransas Bay, and rum runners slipped through waterways as deftly as had blockade runners during the Civil War. Wildcatters took leases on land around Aransas and Nueces Bays. Troops camped again on Corpus Christi Bay while others recuperated in a sanatorium on its North Beach peninsula. Young suburbanites built new homes just west of their hospital, using the recently repaired causeway to cross Nueces Bay.

In August 1919, the city of Corpus Christi persuaded the Army Corps of Engineers to remove a shoal in the bay that had impeded some ships' passage.[28] No closer to a deepwater port than it had been three years earlier, that was the most its leaders could do.

Then the hurricane came. It had its origins in the hot September waters off the Virgin Isles, their moist vapors violently twisting sky-wise even as counterpart winds tore downward, a recurring cycle of heat and chill turning the stale air of the Caribbean into a broad, ever-growing sweep of spiraling thunderclouds and rain. The storm mass moved westward, borne by trade winds, increasingly turbulent pressure systems, and a southern-stretching Bermuda High until it crossed into the Bahamas. Over even hotter waters it gained strength, its force so centered it sucked seas onward. By the time the gale cleared the Florida Straits, it had destroyed ten ships, flooded Havana, and devastated Key West. Then, absent of land and unfettered, it prowled summer waters, their heat reviving even more the spiraling whorl within. No ships survived to report its power, and weather stations along the entire coastline worried.[29] By the thirteenth, even the *New York Times* commented on the uncertainty, and all the *San Antonio Express* could state was that it was "still somewhere in the Gulf of Mexico."[30] Unimpeded and violent, the maelstrom raged closer to New Orleans than the Yucatán, then, almost abruptly, veered westward. A vast, surging, tightly coiled vortex of waves, water, and wind, it approached Padre Isle.

Almost frantically, like forest animals from flames, sea life fled from it. Game fish crowded the pass, seeking refuge and nutrients before the oncoming surge. Flounders gathered in estuaries, so numerous they could be forked with kitchen utensils. Crabs teemed among bayfront piers, sharing space and shelter with shellfish and mussels. Birds filled the air, their numbers increasing by the hour as they retreated inland.[31] Far less wary were humans, still reliant on weather advisories as uncertain as the trajectory of the hurricane itself. By Saturday night authorities had tentatively sited the disturbance as "apparently central in [the] Gulf south of Galveston" but could not locate it more definitely.[32]

Certain factors hinted at the proximity of the storm: wind directly from the north, unusual that early in the fall, and a tide abnormally high. But it took a midnight phone call from an agent in Port Aransas to sufficiently alarm mainland residents: winds were measuring up to twenty-five miles per hour, tides were over four feet, and waters were roiling in the Harbor Island Basin.[33]

It was there and along the edges of Padre, Mustang, and St. Joseph Isles that the swollen billows of seawater burst near Sunday noon. Great surges laden with broken ships, thrust-up bottom mud, and dislodged benthos, they swooped onto shores already washed out by rain. For the cowhands still on Padre Island, the only recourse was the highest point on the isle: "We lay there flat and watched all that water," one recalled. "It was like the whole ocean crashing in."[34] For a pleasure party to the north of them—a charter boat of tourists camping on the outlet between Padre and Mustang Islands—the waves were so shattering that their shelter collapsed and they had to flee to high dunes. One, separated from the others, watched them running to one ridge until it swept away, then darting to another. "The hills melted like snowflakes as the great waves would strike them."[35]

The dunes on the north side of Mustang were no more solid. Many Port Aransas families stayed home Saturday night, only to find ground floors submerged at dawn and foundations soon wobbling in the floods. Overcome by panic, some had to be physically restrained by rescuers; others hastened to the hills, "passing from . . . dune to . . . dune . . . [with] the waves . . . eating into the sand which dissolved like sugar." Even the oil tanks near the town and on Harbor Island fell victim to the surge, half-empty ones overturned and tossing wildly, their contents spilling out onto already laden seas. Basin boats crashed against wharf pilings and heaved onto each other, trestles flew upward and out, supports cratered and piers splintered, and cotton bales joined lumber and planks roaring westward with the waves.[36]

On St. Joseph Island, humans fled, leaving 6,400 head of Hereford cattle penned against oncoming seas.[37]

On the mainland, people of Aransas Pass awoke to an incoming flood. But "it wasn't until the water went over the railroad tracks," one admitted later, "that the seriousness of the situation was realized." Now too aware of danger, residents headed for high ground, "struggling men, women, and children . . . escap[ing] . . . from wrecks of the first buildings on the shore line." As they piled onto more stable shelters, sounds carried over the thunder—wrenching timbers, breaking glass, peeling rooftops, careening railcars, and screams, sudden and short.[38]

Rockport suffered, too, knowing by Saturday evening that a storm was coming and just as certainly that "there was nothing anyone could do about it. No one was fool enough to . . . risk a frail buggy or a balky motor car on open flat land." By Sunday morning water was over the first road, the bay was rising, and the wind aroar. Families tied themselves to clotheslines to struggle across streets, couples straddled horses to reach shelter, brothers breached homes to rescue old ones. Many fled to live oak groves, others to brushy areas on high ground. By the time the surge swept the bay into town, some who had stayed homebound—the Franks family, so sure God would protect them; the Littles, restaurant-proud—had died, drowned in the very streets they had walked a day earlier.[39]

People in Corpus Christi had been more blasé about the coming tempest, even after the midnight call from Port Aransas. Sunday morning dawned wet, windy, and with the bay rising as forecast, but word was out that the storm was small and posed no immediate danger. By midmorn, however, hurricane flags had gone up again, wind from the north was increasingly wild, and some doubt had risen. Surely the city had proved its solidity; virtually untouched by the 1916 storm, it still sheltered in its coastal cuplike eddy, protected by the blackland plains behind it and the bay beside it. But gusts of over seventy miles an hour knocking awry weather tools belied old complacencies, and a sudden shift in bay waters, from bank-full to overflowing, was shocking. It was not, however, until the floodwaters rose from eighteen inches to five feet in less than an hour that anguished uncertainty

Hurricanes

became panic. Corpus Christi Bay had turned against its people—early northerly winds had pushed its waters southward into Laguna Madre and then, with the hurricane's approach, thrust them shoreward, engulfing the business section and North Beach and augmenting surges from the sea.[40] Almost within minutes Nueces and Corpus Christi Bays became torrents plunging inland, battering and beating and mauling everything in their way. Families crowded rooftops, only to feel them collapse beneath them; fathers clung to babies, only to lose them to the waves; women clung to debris, only to fall unconscious and sink. A nun and her charges died in the disintegrating Spohn sanatorium, and Teddy Fuller's aunt drowned, burdened by her own sodden clothes. Rest-camp soldiers pulled victims to safety but lost comrades in the catastrophic waves.[41]

By the time the hurricane tore across mid–Padre Isle, it had denuded North Beach of all buildings save three, destroyed the Municipal Wharf, and swept a half block of structures and all the timber and planks and logs and oil from the barrier islands onto an inundated business district. Seven miles of S.A.A.P. track were down, the railway trestle connecting with Portland was crippled, and the causeway was once again in pieces, some of its concrete girders thrown more than a hundred feet westward.[42]

The death toll was staggering. Over a hundred bodies washed ashore on the north side of Nueces Bay, ripped naked, crushed, and covered with

Downtown Corpus Christi streets after the 1919 hurricane. Courtesy of Jim Moloney.

oil. More surfaced within the city itself, pummeled onto streets layered with pilings and barrels and bales. Some sank deep into bay-floor mud, buried under waters still coursing inland, while others, tangled in debris, drifted westward, landing eventually on distant riverbanks. Not all of the nearly three hundred who drowned were of Corpus Christi; the storm killed eleven of Rockport's citizens, seven from Port Aransas, and five from Aransas Pass. But the majority were Corpus folk, ravaged and mutilated by a hurricane that had grown deadlier the closer it came. Without the dunes and brush and live oaks that saved most islanders, they died in droves.[43]

Except for those who lived on the bluff. Riding out a storm that seemed even less wind driven than that of 1916, they awoke to a bayfront below roiling with broken bits of foundations and piers and people. Pelted by an early dawn rain, they stood in almost uncanny awe, gazing down. "It looked like a tidal wave had come," young Alclair Mays Pleasant remembered, "dead bodies floating, debris from buildings . . ."[44]

The high brushy ridge saved those residents from harm only to thrust them into retrieval of the most dismal kind. Donning boots and grabbing boats, men pushed into swollen waters, lifting survivors from the wreckage and transporting bodies to the county courthouse. Women crowded kitchens and storage rooms, herding sons into rescue service and daughters to cooking stations. Politicians wired for help, desperate for the military troops and Red Cross workers who converged on Corpus Christi almost immediately.[45]

Corpses began to be identified, debris began to be cleared, charity began to flow in—and complaints began to be heard. A county judge criticized the dilatory removal of the "lumber and other property which has drifted ashore along the Nueces Bay" five days after the storm. A Staples Street doctor badgered the city for $500 in damage compensation six days after the storm. Aransas Pass challenged the Relief Committee's monetary dispersion policies thirteen days after the storm. Over $104,000 "is going into Corpus Christi for the whole stricken area," W. E. Tedford complained to authorities, although "we have received only two thousand dollars."[46]

It was a valid complaint. Rockport and Aransas Pass had received very little outside aid by way of the Corpus Christi Relief Committee, Port Aransas even less. Yet they needed it. Located near an island where the hurricane's last victim would eventually be found, Aransas Pass had lost

Hurricanes

its city wharf, a swimming pavilion, and the two-story headquarters of the Aransas Pass Terminal Railway. Its Live Oak Ridge tank had toppled and spilled oil into the oncoming surge. Its rail line to Harbor Island Basin had cratered, as had basin wharves and port-side storage vats. The north jetty at the pass had ruptured, as had the dike atop St. Joseph Isle.[47]

Rockport was particularly affected, with "wreckage everywhere," Colonel W. D. Cope reported; "the First National Bank Building is destroyed

Coastal Bend towns in the aftermath of the 1919 hurricane. *Above*, Aransas Pass, courtesy of Corpus Christi Public Libraries; *below*, Rockport, courtesy of Texas Maritime Museum.

as are hundreds of others." Its whole waterfront was devastated and its S.A.A.P. tracks to Aransas Pass were completely gone. Bailey's Pavilion had foundered, and the dance floor had beached in what had been a vacant lot. Parlor stools and fabric bolts dotted the shoreline; pianos leaned brokenly against fences; homes tottered atop tilted foundations. An iron vault lay overturned on its face, scraped across a floor by the waves.[48]

Survivors stepped cautiously, dodging broken glass on the streets and pushing off snakes carried in with the surge. "Water snakes and marsh snakes," one resident remembered, "garter snakes and ribbon snakes . . . coachwhips and racers . . . vipers and rattlers." They clustered, clinging to shirts of rescuers inland and attacking careless workers on the isle. Nor were there snakes only—crabs had flowed into bedrooms with the waves, and dead fish and turtles crowded shores. Drowned cattle lay stiffening, their limbs pointed outward; soon gulls would be pecking at their eyes. Mosquitoes were out in full force, with flies covering bodies like blankets of soot. And the stench, a survivor remembered—"wet lumber, moldy dry goods . . . slime, mud, dead animals, decayed fruits . . . burst septic tanks"—the stench "was terrible"; it pervaded all.[49]

Yet for some, the outlook appeared positive; not all the hurricane wreaked was bad. Physically, landforms had changed. The shoreline on Corpus Christi Bay had been heightened by sands brought in by the waves; some claimed it actually "rivaled the famous beach at Atlantic City." And waterways grew deeper. Padre Island's Packery Channel was now nearly twenty feet deep. The strait between the jetties at Aransas Pass had been gouged even deeper, the force of the tide dredging it "to a far greater depth than ever known before." Financially, creditors were generous. The government absorbed the loss of damaged ships at Heldenfels Shipyard, and business owners won extensions on overdue loans, occasionally getting replacement goods free of charge. It was no mean accomplishment that a mere twelve days after the hurricane ripped through the area, a news reporter could state that "the general outlook of conditions Thursday was considered to be the best since the fatal fourteenth of September."[50]

But there were those who wanted more than a return to status quo. They wanted a jolt into the future, and none were more determined to achieve it than Roy Miller and Robert Kleberg—for them the future was still a deepwater port.

PART V. EXPANSION DAYS

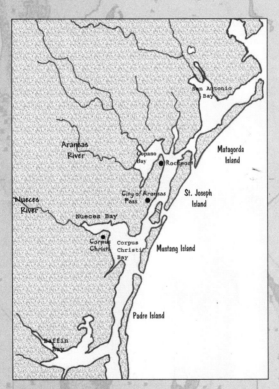

Cities in contention for deepwater port, 1919–1922.

CHAPTER 16

Hall's Bayou Enhanced

Working with city leaders, Miller and Kleberg had a firm foundation on which to build. The horror of the September hurricane had made news all over the nation; there was no major metropolis that did not know of the damages and few that had not contributed some funding for relief. Institutions like the Red Cross and the US Army had involved themselves in helping the area, as well as the Texas National Guard, the Salvation Army, the Masons, and the Knights of Columbus. Congressman Carlos Bee was to visit soon, promising federal aid for more secure protection. What better time could there be to extend such concern into a bid for "an ocean port . . . at this city"?[1]

With gratitude and not a little bravado, Corpus Christi's establishment welcomed Bee, embraced his support, and added a proviso—with government funding they would not only build a safe and adequate seawall but they would build it "along the bay front of the city and . . . into Nueces Bay which would be converted into one of the finest harbors of the whole United States." Government-provided protection, they implied, compelled the building of their own deepwater port. Experienced in political gamesmanship, Bee vowed to consult with Port Aransas and Rockport before committing to any such plan, but he did wire the Army Corps of Engineers to activate a "survey of the channel from Port Aransas to Corpus Christi" that had been commissioned two years earlier.[2] Thus the process began.

Relying on an accumulation of state-remitted ad valorem taxes for the purpose of bayfront protection, city council members commissioned plans for a fourteen-foot-high seawall, opted instead for a six-foot-high breakwater, then set their sights on persuading the military to reverse its 1910 decision setting Harbor Island as the area's premier deepwater port. "The storm has ruined Port Aransas as a harbor," Roy Miller reminded citizens,

Breakwater being built in Corpus Christi Bay. Courtesy of Murphy Givens.

"and has demonstrated [its] impracticality. . . . Now is the chance to show the government that it should establish a deep water harbor here."[3]

Building the breakwater on the city's own initiative was one way "to constitute a favorable argument in favor of Corpus Christi," one reporter noted. Selling its advantages—high bluff lands, extensive railroad connections, solid banking facilities—helped, and getting the state to commit twenty-five more years of ad valorem taxes "for port purposes" enhanced its case. But it was the activities of two groups—the Corpus Christi Deep Water Committee and the South and West Texas Deep Water Harbor Association—that all but cinched Corpus Christi as leader in harbor competition. Ramrodded by Roy Miller and Robert Kleberg, these advocates even convinced the San Antonio Chamber of Commerce to support the Corpus claim.[4]

The next step was even more effective than self-built breakwaters—Nueces County residents petitioned to create a local government agency, a political subdivision of the state, that would build the actual facilities of a port. The new navigation district would not only cover construction but would also, with government approval, dig a conduit from Hall's Bayou (a low area on Corpus Christi's shoreline) inland to a turning basin adjacent to Nueces Bay, construct within it wharves and docks, and align those with warehouses and train tracks leading out of the city. Although bay channel dredging and maintenance would fall to the federal government, this port would be owned and operated solely by its community—no private interests could monopolize. The army's role would be to dredge the Corpus Christi Channel to a twenty-five-foot depth from Aransas Pass jetties all the way to Hall's Bayou—and recommend that the city be the South's new deepwater port.[5]

Organizations, letters, government bodies, statements of persuasion—all contributed to the Corpus Christi sell. But the personal touch was necessary as well, and Kleberg and Miller had that well in hand. They not only lobbied influential members of the Rivers and Harbors Committee but also kept Major L. M. Adams, engineer in charge of Texas waterways, abreast of developments. It was Adams whom the Deep Water Committee notified about engaging an industrial expert to compile comparison data regarding a Corpus Christi port. It was Adams who knew of the detailed plans for wharfage, warehouses, and terminals being developed by an outside consultant. It was to him, as district engineer, that the city council confirmed that any state remittances "shall not be retained or paid to the City of Corpus Christi until . . . [it] shall have been designated as a deep water port." And it was Adams who would coordinate the forthcoming Corps of Engineers survey teams and prepare their reports to Congress. The major was ready. He acknowledged the need to decide "once and for all whether a safe, adequate and better port to meet the real present and perspective needs of southwest Texas is justified." He was also aware of the public affairs aspect of such an obligation: that "if any amplification of existing projects is to be undertaken, it should be with reasonable assurance that the work . . . will receive the hearty cooperation and support . . . of Southwest Texas generally." But foremost in importance to the major was support from "the communities immediately contiguous" to the "existing projects"—and that was in jeopardy.[6]

For the barrier isles had rebounded. True, hurricane-deepened Packery Channel inlet had shoaled up once more, but Padre Island was swiftly regaining the grassy surface of which it had been stripped, and shorebirds had returned. Sunken lumber and caved-in hulls were providing new homes for mussels and shipworms, and whole communities of sea life were sheltering in submerged trestles strewn across the bays. Not only was nature restoring itself; shipping interests were as well. By December 1919, the long channel between Aransas Pass and Port Aransas had been cleared out and deepened, and the government vessel *Comstock* was due to dredge the main channel through the pass.[7] By June of the next year, the barge *Providence* had been towed to the Harbor Basin wharf with fifteen thousand barrels of Mexican crude oil to be transported inland. "This is the first deep water cargo discharged at Port Aransas since the storm of last September," an observer noted, but it would not be the last. "While Port Aransas is in position to handle oil shipments, the terminal rail facilities were so badly demoralized," Captain Ellisor warned, "that it would be impossible to handle a general cargo there at present . . . but this section of the coast can come back into deep water traffic if . . . the people will go after the improvements that must be made."[8]

That was a basic question facing the US Army engineers as they surveyed the region: If a deepwater port was still viable, were people of the coast willing to go after improvements to restore Harbor Island Basin—or would they, like those of Corpus Christi, try to get their own port? It seemed the owners of Harbor Island Basin were ambivalent. The Aransas Pass Terminal Railway Company did repair railroad trestles from Harbor Basin to the mainland, and the Aransas Pass Channel and Dock Company carefully rebuilt wharves for ships coming through the pass. But demolished "sheds and storehouses" were still in shambles four years after the storm.[9] The only proposal Adams's team received to make the basin a "safe, adequate and better port" came from the district engineer himself: to raise facilities twelve feet above sea level, to build a seawall around the harbor side of the island, and to erect a concrete causeway connecting rail lines to the mainland. Port owners, however, did "not consider that works of this kind are needed," Adams observed, and they offered only lukewarm support for an improved channel and turning basin.[10]

In contrast, Aransas Pass residents wanted deepwater designation. They promised an offshore port, with channel and turning basin about one thousand feet in front of town, protected by a piling-secured levee. The basin would include a wharf one thousand feet long, "with a warehouse and track and road connection to the shore." The channel itself, connecting the turning basin to the pass, needed to be the standard twenty-five feet deep; it was its width that led to contention. District engineer Adams maintained it should be two hundred feet wide; his superior, division engineer Colonel H. C. Newcomer, recommended it be narrower. The two clashed again in their estimates of overall expenses, Adams estimating $1,637,000, Newcomer only $1,150,000. Yet the advantage lay with Aransas Pass. Although it had lost the longtime backing of San Antonio, the town had the continued support of its single rail line, the S.A.A.P. Its willingness to provide a safe, adequate, and relatively cheap deepwater harbor gave it a competitive advantage.[11]

Rockport's plan pulled on every insight learned over the years. Its port, located offshore one and a half miles from town near the S.A.A.P. lines, would be protected on the south by a concrete seawall and on the east and north by earthen embankments. The turning basin would be only 1,200 feet square with dock, bulkhead, and two berthing spaces, but provisions were built in for expansion. Rockport's long-used route from the jetties along Lydia Ann Channel would be dredged to the norm, twenty-five feet. In addition, not only was the entire harbor to be elevated fifteen feet above sea level, but the people of Rockport were petitioning to form a navigation district and were also inveigling the state to remit ad valorem taxes from three counties to offset embankment expenses. The overall cost with just limited berths was only $2,032,000.[12]

Corpus Christi's plan included a turning basin 1,200 feet wide and 3,000 feet long at the end of a 25-foot-deep channel cutting in from Hall's Bayou, wharf construction allowing for seven vessels with provision for expansion, completion of breakwaters near the harbor, and a lift bridge carrying S.A.A.P. lines across the entrance inlet. The turning basin was to be cut from land next to Nueces Bay and a levee would be built to protect it from overflowing waters. A requested navigation district would, if all proposals were approved by voters, draw on almost $100,000 of bond-issued funds to facilitate building; city and private interests had already pledged an additional $90,000 toward that end. The army would extend the Corpus Christi Channel from Turtle Cove across the bay, widen it to two hundred feet, and deepen it twenty-five. The total estimated cost was over $5 million.[13]

Corpus Christi won. Financially the odds were against the city. A lineup of federal approximations put Corpus Christi's proposal at least $900,000 over that of its closest rival, Rockport. Estimated costs to local interests showed Corpus Christi exceeding her rivals by more than $2 million. In addition, difficulties common to dredging any open bay would be multiplied by the deepened channel proposed between Hall's Bayou and Aransas Pass. But even accepting how much cheaper a shorter conduit to the town of Aransas Pass would be, and how eager Rockport citizens were to levy bonds even at the risk of reaching their limit, nothing could surpass the benefits offered by the city of Corpus Christi. "It is a thriving city of about 10,000," Brigadier General H. Taylor wrote in 1922 at the conclusion of two surveys, multiple meetings, and a preponderance of reports. Not only was a deepwater port necessary "to this section of the coast of Texas," but Corpus Christi was "the logical place. . . . It is served by four railroads . . . it has the assurance of assistance from commercial and transportation interests, and [it] commands widespread confidence as to its ability in carrying this difficult project to a successful issue. It is factors like these," he added, "that have led to the development of ports far from the sea, despite the expense of excavating and maintaining channels leading thereto." Well-constructed ports "far from the sea," like that of Manchester, in Great Britain, were what Taylor must have had in mind. That forty-mile seaway connecting the city to the Irish Sea, less than thirty years old, was already outstripping seaside ports in total tonnage shipped, while Houston's relatively new Ship Channel, stretching down Buffalo Bayou through Galveston Bay into the Gulf, was turning that city into a leading cotton exporter. Surely dredging a twenty-mile stretch from Aransas Pass to Corpus Christi wharves would facilitate similar success.[14]

But there was one more reason, possibly even weightier than that of inland-port advantage, that turned the army specifically toward Corpus Christi: the city promised a landlocked turning basin. Such an offering would not have been made years earlier, Major Adams admitted in his December 1920 report. "As previously stated," he reminded his superior, "there had been no serious hurricane visitation at this locality for thirty years." People lived in a false atmosphere of security, even forgetting the great storm "of 1886 which completely destroyed Indianola." The

Hall's Bayou Enhanced

devastations of "the serious hurricane of 1916 and the terrible disaster of 1919" demanded radical change, however. Harbor Island, for a start, lost credibility as a deepwater port. Even if reluctantly supported by the Aransas Pass Channel and Dock Company, any improvements on the isle would remain "entirely insufficient. . . . [The] location is, and will continue to be, subject to serious damage by West Indian hurricanes." Although Rockport and Aransas Pass sincerely desired port status, he continued, their suggested sites, each fortified by levees and revetments, remained off-shore. "None of the localities examined can give proper assurance of . . . a safe and adequate harbor . . . except the one adjacent to Corpus Christi. Here the proposed turning basin, 1,000 feet square, will be landlocked and should afford a safe berth to any vessel seeking it . . . on the visitation of a hurricane." Thus did a wind-driven calamity and an inshore-designed plan combine, propelling the city into the future.[15]

Congress's Rivers and Harbors Committee acted on the Board of Engineers' recommendation in May 1922, the bill designating Corpus Christi

Painting of the Port of Corpus Christi, 1924, as it would be upon completion. Courtesy of Murphy Givens.

as "the deep water port on the west gulf coast" went to President Harding five months later, and on September 22, Roy Miller received word that the struggle was over. "Harding Signs Port Bill," the headlines read.[16]

Within five weeks, the navigation district was approved by county voters along with authorization for it to float a million-dollar bond issue to construct port facilities; three days into the New Year citizens approved more bond sales to activate state allocations. Armed now with official power, assured funding, and fifty acres of donated land (offered by gubernatorial candidate W. E. Pope), the newly designated Navigation and Canal commissioners got to work—and totally transformed Colonel Kinney's coast.[17]

Instead of the sunken lowland muddying within minutes of any intermittent rain, a district-commissioned dredger chugged Hall's Bayou into a conduit twenty-five feet deep and two hundred feet wide at its bottom. Sodden and soggy for years, it became a notable channel, extending a tenth of a mile inland and opening into a vast, water-filled basin one thousand feet wide, three times as long, and nearly thirty feet deep. Moreover, piled on the basin's north and west sides was the bayou's own soil, dredged from its bottom and thrust, along with shells and sands, into a fifteen-foot wall. The levee along the basin's south side was even more solid, formed from inland clay and bounded by stone; from it would extend a wharf into the basin, twelve hundred feet long. The most wondrous change, however, was the land between that levee and the northern side of the bluff—layered with dirt pulled from the basin, it had grown fifteen feet in height and comprised one hundred acres on which would be built warehouses, railroad yards, grain elevators, and compresses.[18]

The changes did not stop at the basin. The beach on which Taylor's troops had camped was now ringed by semicircular breakwater structures, two thousand feet out and over twelve thousand feet in length. Covered by rocks unloaded from trains tracking into the water and placed by floating derricks, the barriers were separated from each other by three-hundred-foot gaps, except where the new entrance channel was. There the space extended over four hundred feet, for that was the culmination of the third great transformation on the bay: its channel.[19]

From the aborted dig started before the Civil War to the old Morris and Cummings Cut now almost defunct, Corpus Christi Bay had resisted

Hall's Bayou Enhanced

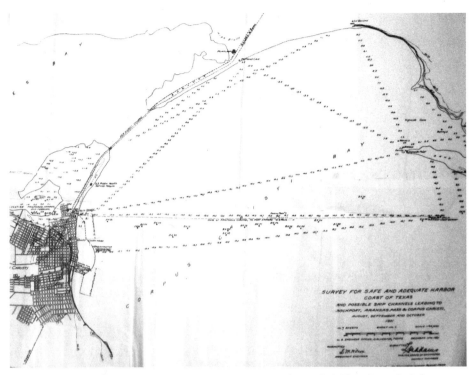

US Army engineers' plan for digging a navigable channel across Corpus Christi Bay. *Survey for Safe and Adequate Harbor Coast of Texas. August, September and October 1921, S. M. Wilcox, Assistant Engineer.*

every attempt to modulate its shallowness. Ranging from three feet near the shore to depths varying between twelve and fifteen feet centrally, it generally accommodated only the lightest of skiffs. Even when it was widened and dredged out of Turtle Cove as the Corpus Christi Channel, first suggested by Rudolph Kleberg, participating ships could bring little but staples to city wharves.[20]

Until the *John Jacobson* arrived at the pass. Huge, sporting the biggest suction pump of its kind, it was one of the largest dredgers the military could commission. Starting down the old channel from the pass, it dug deep, loosening sands, silt, and invertebrates and sucking them into its maw. Continuing westward onto the bay itself, its cutters half circled slowly, clawing down and pumping up even more bay floor as it moved. Finally, filled to satiety, it streamed layer upon layer of mushed muck and sea life outward. But unlike earlier dredgers, the *Jacobson* aimed as it spewed. Soon, mounds of mud began to emerge from the bay floor, growing into hills and finally islands. Primordially created itself, Corpus Christi Bay gave birth to its own landforms. Strategically sited as only engineers could do, the spoil islands lined the developing channel, marking not only its four-hundred-foot width but signaling as well the steady progress of its dredger. By July 18, 1926, "exactly one year, six months, and nine days after the cutter was dropped at harbor," the 20.78-mile-long channel across Corpus Christi Bay joined the breakwater entrance to the Port of Corpus Christi.[21]

CHAPTER 17

Holding Firm

The completion of the port did not mean the end of the Corps of Engineers' interest in the Gulf. Waterways from New Orleans to the Coastal Bend would soon be churning with military-commissioned dredgers, digging the nine-foot-deep, one-hundred-foot-wide Intercoastal Canal approved by the president in early 1927. Additional dredging unrelated to standard maintenance loomed ahead as well, the port so efficient in its first year that supporters were already lobbying Congress for an extra five feet of depth in the channel. Engineers were monitoring the pass as

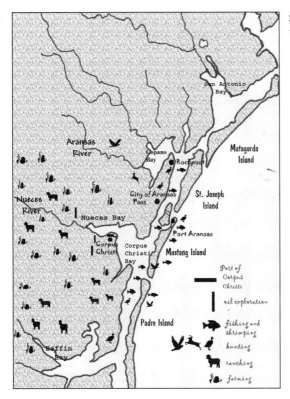

Some area industries in the late 1920s to 1930s.

Oil derrick near the community of Aransas Pass. Courtesy of Jim Moloney.

well, lauding Port Aransas's rapid rebuilding of its public wharf and adding four rubble-rounded spurs to the north jetty to reduce scour even more. Their attention to other Coastal Bend communities had lessened, however, for "the designation of Corpus Christi" as the site for deep water "renders unnecessary any increase in channel depths for Rockport or Aransas Pass."[1] In terms of waterway improvement, at least for the present, these towns were on their own.

In terms of petroleum development, however, they were on a roll. The stalking of the mainland and islands initiated by wildcatters after the Great War led to hope for the town of Aransas Pass. Entrepreneur John Sigmund had drilled four wells there by 1925, and the explosive success of one—"the gas was so strong that it carried the mud three-quarters up the way of the derrick"—indicated significant oil in the immediate vicinity. Even if that proved untrue, Sigmund proposed "to come back and develop the gas," a reporter noted.[2]

It was crude oil from West Texas that gave Port Aransas hope. Pumped four hundred miles across the state to the coast, some of the petroleum poured into the Humble Company's refinery at Ingleside, the Port of Corpus Christi's new adjunct port, to be purified for later shipment. But most had been piped over bay floors to Harbor Island, there to rest in huge storage facilities until discharged into tankers even larger than those that came in the early century. "These giant enterprises have brought new life and new capital to Port Aransas," the *San Antonio Light* observed. "They furnish employment to six hundred men who command a pay roll ranging from $4.50 to $25.00 a day."[3]

A different kind of employment saved Rockport. The large numbers of laborers that had worked at Heldenfels Shipyard had dispersed, and the shipyard itself had diminished in importance as the century progressed. But the community's devotion to its bay never died, and in 1925 it set up its own navigation district. Within a year a new harbor had been created offshore, and Rockport revitalized one of its oldest industries: shrimping.[4]

Among all the marine life harvested since Texans began commercially fishing for food, *Penaeus setiferus*, or white shrimp, seemed one of the least affected. Fearful that excessive hauls would depopulate a species already notorious for its short life span, the state, by the early twentieth century, had restricted shrimpers' activity to the use of seines and cast nets only. The limited length of the scoops and of the arms of their handlers meant that such gathering could occur only in bays and only for full-grown shrimp. Those less mature had to be thrown back. This gave the juvenile decapods, spawned on the sea mud of the Gulf and swept into the pass by currents, time to thrive in the waters of Laguna Madre and its estuaries. By the time they matured—up to seven inches in length, with elongated heads, thoraxes, and large muscular tails for propulsion—many had already swum through the pass and back to breeding grounds in the Gulf, out of harvesting territory. In fact, by 1908, the shrimp yield in the fishery industry had dropped over 80 percent from its nineteenth-century high. Restrictions on its capture, the shallow depths covered by seine nets, and problems with rapid decay kept this species relatively untouched in the years preceding the Great War.[5]

Time and technology changed that, however. In 1920 the state Game, Fish, and Oyster commissioner gained the power to license otter trawls for shrimping, and within two years the oval net bags became a fixture of the Coastal Bend. Far larger than seines—otter trawls, now known as shrimp trawls, extended over fifty feet with mesh up to two inches wide—they were considerably heavier as well. Besides the yards of net bagging, which came to an open funnel at one end, the bottom rim of the net was weighted with lead balls to keep it down. On the upper edge of the bag were floats to keep that part of the apparatus on the sea surface. Attached to each end of the wings of the bag was a board, iron-strapped on its bottom and

Shrimp boat traversing waters in the Coastal Bend. Courtesy of Texas Maritime Museum.

attached to a ship's towline on its top by a bridle-like fastener. Pulled by both towlines behind a power-driven boat at three miles per hour with the funnel end fastened securely, the trawl could scrape virtually every living organism off seafloor and bay bottom.[6] Shrimp were the primary object, and they came in every size and shape, many young and still growing in estuaries and lagoons, others full-fledged and already in the Gulf. For, with restrictions on the type of netting removed, shrimpers could venture into open waters and trawl for adults.

And many did. Some who left Aransas and Corpus Christi Bays—and many who left Matagorda and Galveston Bays—were in large vessels, five tons or more, with huge power winches that reeled in heavy-laden trawls and emptied the shellfish on deck, to be culled and dumped into a refrigerated hold. But the majority of those headquartered in Rockport and adjacent towns in the 1920s powered smaller craft, about twenty-five feet long, that drew only three to four feet in depth. Their captains knew the inlets and coves and estuaries where they had been raised; they understood the seasons and the best times to fish; they understood the need to throw out small cast nets before committing large trawls. Most of all, they understood the short life span of their quarry and worried, as did members of the Game Commission, about depleting their stock.[7]

Others worried, too, not just about overfishing the crustaceans but about destroying their young. "Tons of these small shrimp, smothered and killed in the nets of the trawls were dumped back into the bays . . . a total loss to the commercial fishermen and to the future use of the fishery," one authority commented. Microorganisms and plankton were swept up too and destroyed, while spilled fuel and garbage followed some vessels. Measures were taken to improve the industry, and the size of the individual shrimp became important once again; some bays were permanently closed. But restrictions did little to curtail the burgeoning growth of shrimping, and by 1936 there were twenty-nine trawls operating in Aransas Bay, thirty-one in Corpus Christi Bay, and significant shipping and canning facilities all along the Coastal Bend.[8]

Thus by the 1930s, the barriers and their estuaries and inlets were more than mere shipping channels; they had become commercial food sources. But to locals who had always fished these waterways, harvesting them meant far more than trawling the sea with nets; it meant enjoying a personal bond with nature. Children fished in Copano Bay while young, clutching cane poles cut small for their hands. Mothers kept toddlers close while dipping for crabs near Packery Channel. Teenagers trailed the shores of Laguna Madre, netting the same species of black drum and redfish as had their Karankawa predecessors. Entrepreneurs leased out yachts, promising "bait, tackle, and meals which may be had on board, if desired." More frugal adults rented rowboats for less than two dollars a day, attached motors to them, and prowled bays, seeking their deep holes to fish. And fish they found: kingfish and mackerel, skipjack and drum, mullet and carp. Men filled containers with redfish and boys sold trout for hotel cuisines. Stories flowed—of boats so full they had to be carried over the flatlands, and of oyster reefs so vast they would shear the pin off a motor.[9]

In their midst, sports fishermen from afar thrived. They haunted jetties looking for silver trout and red snappers; they cruised near wrecks, hoping for bluefish and pompano; they hired experts to take them to groupers and perch. But many were dedicated tarpon hunters who had mastered their prey in the southwestern bays of Florida. Eager for new challenges, they became seasonal visitors, willing to brave the difficulties of travel to Aransas Pass and its islands. In 1926 their trek became easier; the little railway that had brought mainland workers to Harbor Island jobs had

Postcard advertising the Tarpon Rodeo at Port Aransas. Courtesy of Corpus Christi Public Libraries.

morphed into a full-fledged transportation system. Visitors could now drive their vehicles onto flatcars, be carried to Harbor Island, and then be offloaded onto ferries to Port Aransas. Access became so easy that a new Coastal Bend industry began to bloom: tourism. Women converted old homes into hotels, old-timers started hunt clubs, organizations initiated summer camps. Most passionate of all were the fishermen, so dedicated to tarpon that they eventually pushed Port Aransas into hosting the most prestigious marine competition in the state. About six years after the dedication of the Port of Corpus Christi, area guides began a local competition termed a fishing rodeo. Within a year, participation was public, and almost immediately the Tarpon Rodeo became a highlight of the state. Sportsmen competing to land the most "spectacular jumper" of the sea came to the little town in droves, many steering down a new asphalt causeway.[10] Others

Holding Firm

boated to Mustang Isle to compete in the event, and numbers grew. But change was chancy and not always constant. It would take more than fishing contests or oil storage tanks to bring Port Aransas the attention it had enjoyed in the heyday of Harbor Island Basin—so a president came.

Franklin Delano Roosevelt already had strong ties to Texas. His vice president, reelected just the year before in 1936, was John Nance Garner, who had pushed so long for the port and the intracoastal canal. One of Roosevelt's closest oil-rich friends, Sid Richardson, had just recently taken possession of St. Joseph Island and was reviving its cattle heritage along with its wildlife. Roosevelt's son Elliott had fished in Gulf waters before and had become entranced with Aransas Pass and its tarpons. It was he who had suggested Port Aransas to his father, and it was there, in the spring of 1937, that the president arrived.[11]

The barrier isles could not have been more responsive—or uncooperative. The first night of Roosevelt's stay saw a storm hit "that would blow the shirt off your back," an observer noted. Tarpon surfaced near the president's craft two days later, but the one Roosevelt caught, a splashy jumper

President Roosevelt reeling in his prize tarpon. Courtesy of Texas Maritime Museum.

that soared, twisted, and turned in the air exactly as promised, tore off the hook and flashed away into the sea.[12] Commercial harvesting got into the picture, a Gulf shrimp trawler so enthusiastic to see him he was showered with a fresh haul. "The president . . . [just sat] there grinning happily, shrimp in his lap and all around him. . . . We had shrimp for the rest of the cruise," a shipmate reported. Hopper-type dredges made their entrance, muddying up Gulf waters so much that Roosevelt and his fishing companions had to head elsewhere. Even tiny, silent, and semisecluded St. Joseph Island, now nicknamed St. Jo, became part of his experience, as Richardson used his cattle chute to roll his old friend from boat to dock.[13]

But the culmination of the president's trip—and the reason he went to Port Aransas in the first place—was his tarpon fight the last Saturday of his stay. Unsuccessful all day, he and his guide, Barney Farley, were heading back to shore when Roosevelt cried out, "Barney, I think I'm hooked on a rock." He wasn't. "A few seconds later," Farley remembered, "a huge tarpon exploded into the air. The rod bent in a half circle as the tarpon hit the water. . . . The fish jumped again . . . the reel singing." Roosevelt fought that tarpon for an hour and twenty minutes, periodically exclaiming, "This is wonderful, Barney. This is wonderful." By the time he landed the fish, the boat was over two miles from its starting point. Roosevelt was elated, taken for a wild man-versus-nature spin he had never experienced before—nor ever would again. The barrier isles had given him their best.[14]

CHAPTER 18

Hawsepipers

What the islands had not given the commander in chief that spring was a cruise between the very jetties his Army Corps of Engineers had built. Had Roosevelt taken the trip, it may have been a hawsepiper who guided him through. Nautically, a hawsepipe was the hole in the hull through which an anchor chain threaded; it was the tube through which all that held the ship steady on the ocean floor connected with all that ran it on deck. The hawsepipe was a vessel's conduit of support. In mariners' terms, a hawsepiper was the same: a person who started from the bottom as an ordinary merchant seaman and steadily worked his way up the pipe, "getting . . . time . . . training" and experience enough to keep his ship steady, master his craft—and pilot the pass.[1]

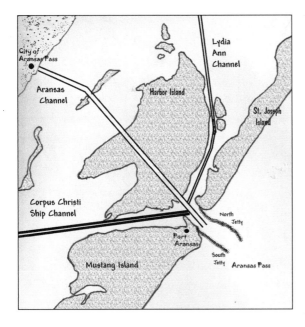

Routes through Aransas Pass by the late 1920s. Based roughly on *Map of the Gateway to the Southwest and the Valley of the Nueces*, Port Aransas Museum.

Tugboat nudging a ship out of the Port of Corpus Christi. Courtesy of Murphy Givens.

The process was slow, time consuming, and often tedious, especially if one started as a grunt on a barge. Even the *Colonel Keith*, a wide, flat-bottomed keelboat-like vessel with an eight-hundred-ton load capacity, could seem a slug. Entering the Port of Corpus Christi on September 8, 1926, it occasioned much bell ringing and joy, townspeople excited to see the first commercial craft move between the supports of the new harbor's bridge. But it was empty. Like barges everywhere, the *Colonel Keith* was a passive instrument, transformed to activity only when loaded with cargo and propelled externally, from outside the boat itself.[2]

It was being able to propel externally, to steer inert vessels through "sudden, unexpected currents," that had decades earlier inspired islanders like Tom Mercer—a man with a lifelong dedication to the sea. Growing up on Mustang Isle, studying its tides and currents, catching its fish and stalking its birds, he had, like his grandfather, seemed to become almost one with the island. So crewing on tugboats was natural; he learned to set fenders over the bow, stoke boilers with oil, maintain towlines off the stern, and relay radioed messages to the master. In time, he captained his own push boats, maneuvering vessels between the jetties of Galveston harbor or into the Port of Houston. But always and ever, Mercer was a pilot, so reputable the army used him to refute Haupt's insistence that his single reaction breakwater had created a navigable pass. Younger men like Clyde Armstrong and Carl Bromley followed him, working their way up the hawsepipe and anticipating the day when they could push barges so big they looked like submarines, or ease four-hundred-foot freighters into a port slip. But most of all, like Mercer, they took the skills they mastered as tugmen with them every day they served as bar pilots.[3]

They had to. It was not enough to complete the training and certification still demanded by the state, or to comprehend the currents within the turning basin expected of tugboat operators. Water movements all along the channel and the pass had to become basic to an Aransas pilot's existence. Once accepting a summons to guide, he also had to note wind speed and direction, determine wave strength, and ascertain sea levels. Tidal movement, though not extreme, was consistent, but tidal surges were as disruptive to commercial shipping in the thirties as they had been at Fort Semmes seventy years earlier. Consequently, knowing when they occurred was vital. Checking air density and pressure was necessary too, as was recording temperature and monitoring weather reports.[4]

When the summons came, however, it was relayed by a radioman now, not by flags at mast as before. By the thirties every military ship and ocean liner possessed a wireless receiver, as did many tugboat and commercial fishers. Knowing about spectra and frequencies was as natural to their operators as operating burgees and pennants had been a generation earlier.[5]

Despite such technology, though, it was a small hand-scrawled card that every pilot carried when he caught the pickup boat taking him to his charge. Written on it for quick reference were the measurements of his assigned vessel. For the SS *Dryden*, he would be boarding a five-thousand-ton freighter, steel hulled, steam driven, and over four hundred feet long. Its cargo, already loaded in port, included three thousand bales of cotton that, depending on water density, might bring the ship within one foot of clearing the original depth as it made its way down the new channel. If his ship were the SS *F.Q. Barstow*, it would be three times the size of the *Dryden* and twice as powerful, carrying 115,000 barrels of oil at near-full

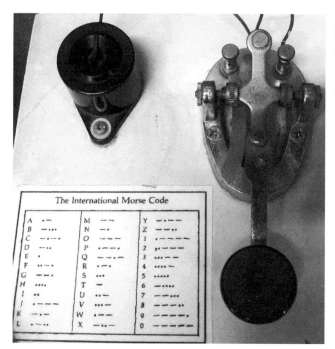

Letters of the alphabet, communicated by Morse code over the telegraph. Courtesy of Texas Maritime Museum.

capacity. It might, however, need shepherding only from Harbor Island Basin, having gone only that far to pump crude from the storage units there. In either case, both ships were huge and each a challenge to board, the pilot jumping from his boat to the Jacob's ladder dangling overboard and climbing its wooden rungs quickly, both vessels matching speed carefully as he rose.[6]

Once on deck, he made for the captain's bridge, calling local events to mind before starting his job: rains that muddied inlets, northers that cut depressions into the bay, wrecks that forced currents askew. Especially noted were displaced channel indicators—buoys shoved awry by barges or shore ranges down. Coordinating range markers was as imperative to modern pilots as it had been to the early Mercers. "You get them lined up, you're OK," one confided. They'll put you "right up the middle . . . of the channel. . . . The deep-drafted ships, they want to stay in deep water."[7]

But the greatest challenge—the problem that made sunken obstacles and whirling currents almost inconsequential—was the 1926 bascule bridge. Built primarily to facilitate railroad crossings and designed as a single-lever lift bridge, the structure spanned the harbor entrance with the widest breadth the city of Corus Christi would fund: ninety-seven feet. As freighters got bigger, and as tankers grew even wider, the free space between ship sides and bridge columns grew narrower. So pilots learned to "thread the needle," ease their hulking boats between supports frighteningly close on both sides. Their goal was to avoid a collision, which could cost the city as much as the bridge's original expense. But accidents happened, and the bascule bridge remained a man-made hindrance for every pilot who brought a vessel through.[8]

Keeping everything in mind, the pilot stood at the wheelhouse window, gauging distances and telling the helmsmen how many degrees to turn the vessel—"port ten" meaning ten degrees to the right, "starboard five," a slighter turn to the left—in order to traverse the channel. The ship's master stood alongside, always in charge and always alert. But for the time it took

Freighters waiting to be loaded in the turning basin, Port of Corpus Christi. Courtesy of Murphy Givens.

to convey a freighter or a tanker into home harbor or out to the Gulf, its safety was in the hands of a Port Aransas pilot.[9]

At the onset of the channel lay the Aransas Pass, now so wide and deep its jetties almost cushioned a ship's passage through. Its pilots and tugmen—hawsepipers some, watermen all—appreciated its dimensions, favoring any port initiative to deepen the conduit and widen its route.[10] Commercial traffic had become as much a part of the pass as the birds and the marine life it had always engendered. Whatever maintained them all kept the barriers thriving.

CHAPTER 19

Henry's Goodbye

That is, to a certain extent. Certainly, the methodical dredgers that removed silt from the pass and scooped out the new channel destroyed innumerable bay organisms. Their incursions affected barriers' sea-grass meadows and estuary nurseries as well, churning up sediments for weeks and blocking off much-needed sunlight.[1]

Currents themselves altered. New inlets, like Ropes's aborted channel through Mustang Island or those slashed through sands by hurricanes, threw longshore drifts off temporarily. The stone-bound jetties that secured

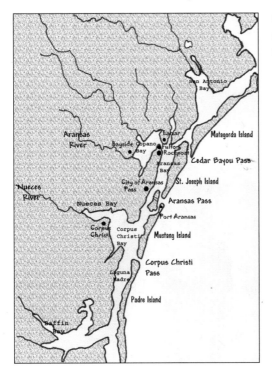

Waterways and communities along the Coastal Bend by the mid-1930s.

the north and south ends of Mustang and St. Joseph Isles, so laboriously attempted and finally built, affected the drifts permanently.

The islands themselves had changed since men had first tried to control them, as had the forces that made them. No longer as vulnerable to Gulf streams that eroded one end and accreted another, they had stopped their southward shifts to stand almost dormant. Sediment still washed downstream and clung to their bases, but much shot outward through the pass, scouring the seafloor as it flowed.

Surfaces were different, too, as business on barriers grew; marshes were drained and tidal flats converted into town lots and residential neighborhoods. Dunes dissolved along with the purslane and morning glories that decorated their slopes; black mangroves and marshes spread on Harbor Island.[2]

Marine life within the pass and its barriers had undergone tumultuous changes. Large reptiles like the green sea turtle were virtually gone, only to emerge infrequently as public barbecue treats. Cold-weather fish kills continued to occur, but their toll was topped by engineers' dynamite blasts. Dolphins occasionally played in the bays, but shrimp, too easily racked up by large trawlers, were now in the process of disappearing.

The jetty, seafloor based and bulwark of life. Courtesy of Jim Moloney.

Pelicans and gulls of the coast. Courtesy of Texas Maritime Museum.

Cormorants atop the Aransas Pass south jetty. Courtesy of Jim Moloney.

Corpus Christi Bay. Courtesy of Jim Moloney.

Deprived of their larvae for food, smaller fish were also feeding elsewhere, their predators—ling, sailfish, and skipjacks—following.[3] The days of standing on coastlines and hauling in game fish were ceasing. It was no accident that President Roosevelt's great tarpon fight happened beyond the jetties in the Gulf.

Even the birds were starting to flee the barriers, at least a few. Whooping cranes, which wintered nearby, were arriving in smaller numbers, and dredging's destruction of shoalgrass meadows was threatening redhead migration as well.[4]

The islands had changed, as had the bays and their shorelines, while the pass's immobility accommodated vaster trade. But, however altered, they were still thriving.

Bay-bottom invertebrates were unquestionably dying by the thousands in the wake of big dredgers, but other benthos simply burrowed deeper, creating tiny submarine communities where they dug. Aransas Pass jetties became inadvertent homesteads, harboring Portuguese man-of-wars and "wing-footed" mollusks. Tiger beetles sheltered in their slabs, sargassum fish and pipefish swam among their gulfweeds, and sea horses and starfish clustered in their base. Brown pelicans used them for landing fields. The bays brought forth new habitats as well, sunken shipwrecks and storm-submerged driftwood now havens for the tiny crustaceans and minute snails on which bigger fish fed. Such "castles in the sea" became fresh fishing grounds for boatmen while their spoil-island sisters attracted cormorants and terns.[5]

And therein lay the magic of the barrier isles, for even as some parts changed and others diminished, the rest continued on, as vibrant and lively as when they first emerged. It was this magic, this spirit, that seemed to imbue humans who settled on them—Patrick Dunn, so caring for his cattle he dug water holes out of Padre Island sand; Roy Miller, so determined to secure deep water in his bay he exploited a hurricane; Robert Mercer, so enrapt with ships at the pass he created a dynasty to pilot them through. And the spirit spread to others only transient within the isles, like Elihu Ropes and Colonel O. H. Ernst. All seemed to share an energy, a dedication that, consciously or not, embraced the Coastal Bend.

It was a force that had to strengthen in order to survive the next few decades. Intense industrialism, rapid petroleum expansion, and increased population would all affect the bays and the barrier isles.[6] But an innate concern seemed to prevail, and within thirty years after Roosevelt landed his tarpon, the people who lived within the Coastal Bend were starting to protect it. Their efforts have been circumscribed by time and practicality. But they remain true to the barriers, devoted to their estuaries, and, like Colonel Henry as he left Corpus Christi for the last time, transfixed by "the bright waters of the bay . . . look[ing] sweetly as ever."

NOTES

CHAPTER 1

1. James B. Swinehart, "Geology," *Encyclopedia of the Great Plains*, accessed September 25, 2016, http://plainshumanities.unl.edu/encyclopedia/doc/egp.pe.028; Carroll H. Wegemman, *A Guide to the Geology of Rocky Mountain National Park (Colorado)* (National Park Service, 1955) accessed September 25, 2016, http://www.nps.gov/parkhistory/online_books/romo5/wegemann/sec6.htm; Robert A. Ricklis, *The Prehistory of the Texas Coastal Zone: 10,000 Years of Changing Environment and Culture*, Texas beyond History, accessed July 10, 2016, http://www.texasbeyondhistory.net/coast/prehistory/images/intro.html; L. F. Brown and J. L. Brewton, *Environmental Geologic Atlas of the Texas Coastal Zone: Corpus Christi Area* (Austin: University of Texas, Bureau of Economic Geology, 1976), 23–24; John W. Tunnell Jr. and Frank W. Judd, *The Laguna Madre of Texas and Tamaulipas* (College Station: Texas A&M University Press, 2002), 14, 28–29, 127–28; Dag Nummedal, ed., *Sedimentary Processes and Environments along the Louisiana-Texas Coast* (Baton Rouge, LA: Geological Society of America, 1982), 15, 17.

2. Wes Tunnell, unpublished critique, September 23, 2016; *Soil Survey of Padre Island National Seashore, Texas, Special Report* (US Department of Agriculture, Natural Resources Conservation Service, and US Department of the Interior, National Park Service, 2005), 10–11; William R. Carr, "Some Plants of the South Texas Sand Sheet," University of Texas Plant Resources Center, accessed July 12, 2016, http://w3.biosci.utexas.edu/prc/DigFlora/WRC/Carr-SandSheet.html; James S. Aber, "South Texas Coastal Wetlands: Padre Island and Laguna Madre," accessed July 12, 2016, http://academic.emporia.edu/aberjames/wetlands/s_texas/texas.htm (site discontinued; printout in author's possession); Brown and Brewton, *Environmental Geologic Atlas*, 23; Nummedal, *Sedimentary Processes*, 8; Tunnell and Judd, *Laguna Madre*, 7, 35–36.

3. A norther is a frigid gale from the north, formed during the winter by a vigorous outbreak of continental polar air behind a cold front. See Dictionary.com, accessed January 23, 2019, https://www.dictionary.com. A hurricane is a "rotating, organized system of clouds and thunderstorms that originates over tropical or

subtropical waters and has closed, low-level circulation." See "How Do Hurricanes Form?," National Ocean Service, NOAA, accessed March 24, 2019, https://oceanservice.noaa.gov; Brown and Brewton, *Environmental Geologic Atlas*, 24; *Soil Survey of Padre Island*, 10; Tunnell and Judd, *Laguna Madre*, 18–19, 127–28.

4. *Soil Survey of Padre Island*, 11–12; W. Armstrong Price, "Reduction of Maintenance by Proper Orientation of Ship Channels through Tidal Inlets" (presentation at Second Conference of Coastal Engineers, Houston, 1951; published 1952), 248, 251; Richard L. Watson and E. William Behrens, "Nearshore Surface Currents, Southeastern Texas Gulf Coast," *Contributions in Marine Science* 15 (1970), accessed January 27, 2019, https://texascoastgeology.com/papers/currents.pdf; Tunnell and Judd, *Laguna Madre*, 130.

5. *Soil Survey of Padre Island*, 14–17; Price, "Reduction of Maintenance," 246–49; Ricklis, *Prehistory of the Texas Coastal Zone*; Orrin K. Pilkey, *A Celebration of the World's Barrier Islands* (New York: Columbia University Press, 2003), 47.

6. Tunnell, critique; "Barrier Island Interior Wetlands," Texas A&M AgriLife Extension, accessed July 12, 2016, http://texaswetlands.org/wetland-types/barrier-island-interior-wetlands; *Soil Survey of Padre Island*, 16–17; "Mustang Island State Park," Texas Parks and Wildlife Department, accessed July 12, 2016, https://tpwd.texas.gov/state-parks/mustang-island/nature; Ricklis, *Prehistory of the Texas Coastal Zone*; Tunnell and Judd, *Laguna Madre*, 122, 226–27.

CHAPTER 2

1. Robert A. Ricklis, "Prehistoric and Early Historic People and Environment in the Corpus Christi Bay Area," Coastal Bend Bays and Estuaries Project, accessed December 6, 2009, http://www.cbbep.org/publications/virtuallibrary/ricklis.html; Ricklis, *Prehistory of the Texas Coastal Zone*; "Barrier Island Interior Wetlands."

2. W. W. Newcombe Jr., *The Indians of Texas: From Prehistoric to Modern Times*. (Austin: University of Texas Press, 1961), 68; Ricklis, "Prehistoric and Early Historic People"; Nancy Kenmotsu and Susan Dial, "Native Peoples of the Coastal Prairies and Marshlands in Early Historic Times," Texas beyond History, accessed August 23, 2016, http://www.texasbeyondhistory.net/coast/peoples/; John W. Tunnell Jr., Jean Andrews, Joe C. Barrera, and Fabio Moretzsohn, *Encyclopedia of Texas Seashells: Identification, Ecology, Distribution, and History* (College Station: Texas A&M University Press, 2010), 6, 9; Tunnell and Judd, *Laguna Madre*, 44–46; Carr, "Some Plants of the South Texas Sand Sheet."

3. The crafts were probably waterproofed with animal grease, asphaltum (sea-based oil), and/or pitch. See "Dugout Canoes," Cherokee Heritage Center, accessed August 30, 2016, http://www.cherokeeheritage.org/attractions/dugout-canoe/;

Richard V. Francaviglia, *From Sail to Steam: Four Centuries of Texas Maritime History: 1500–1900* (Austin: University of Texas Press, 1998), 22–26.

4. Cecilia Venable, Terry Palmer, Paul A. Montagna, and Gail Sutton, *Historical Review of the Nueces Estuary in the 20th Century: Final Report for Texas Water Development Board*. (Corpus Christi: Harte Research Institute for Gulf of Mexico Studies, Texas A&M University–Corpus Christi, October 2011), 1; "Barrier Island Interior Wetlands"; Aber, "South Texas Coastal Wetlands"; Ricklis, "Prehistoric and Early Historic People"; *Soil Survey of Padre Island*, 9–11; Tunnell and Judd, *Laguna Madre*, 25–26; Richard Watson, *Geologic Framework of the Eolian Sand Plain and the Central Flats of Laguna Madre and Circulation between Northern and Southern Laguna Madre*, Texas Coastal Geology, July 10, 2008, accessed January 27, 2019, 11, http://texascoastgeology.com/papers/giww_report.pdf; Claudia Kolker, "The Salty Lagoon," *Texas Parks and Wildlife*, July 2003, 55.

5. The two most generally used terms for these tribes are "Carancaguases" or "Karankawas," although other spellings exist. Kenmotsu and Dial, "Native Peoples of the Coastal Prairies"; Glenn Welker, "Karankawa Literature: The Karankawas," Indigenous Peoples Literature, accessed August 23, 2016, http://www.indigenouspeople.net/karankaw.htm; Ricklis, "Prehistoric and Early Historic People"; Robin Varnum, Álvar *Núñez Cabeza de Vaca: American Trailblazer* (Norman: University of Oklahoma Press, 2014), 134.

6. Because St. Joseph Island was not officially renamed San José until 1973, all references will use its original name. See J. Guthrie Ford and Mark Creighton, "Our Bali Hai'i and a Watery Railroad," *PAPHA Newsletter*, March 2010, 4.

7. Compasses and compass cards helped indicate the direction in which the ship was going; log ropes showed speed; and lead lines helped determine water depth. Astrolabes and cross-staffs sighted the sun, which helped approximate distance from the equator in degrees, minutes, and seconds, while nocturnals used the North Star as a base for determining the ship's position at night. See "Navigation of the American Explorers, 15th to 17th Centuries," Penobscot Maritime Museum, accessed August 30, 2016, http://www.penobscotmarinemuseum.org/pbho-1/history-of-navigation/navigation-american-explorers-15th-17th-centuries; "History of the Astrolabe," Astrolabe, accessed August 31, 2016, www.astrolabe.org.

8. For the full story of this expedition, see Varnum, *Álvar Núñez Cabeza de Vaca*, 26–30, 37, 49; or Andrés Reséndez, *A Land So Strange: The Epic Journey of Cabeza de Vaca* (New York: Basic Books, 2007). See also Donald E. Chipman and Robert S. Weddle, "How Historical Myths Are Born . . . and Why They Seldom Die," *Southwestern Historical Quarterly* (January 2013): 232. Distance between Rio Soto la Marina to Tampa Bay, Florida, is 2,167.4 km (1,346.5 miles). See Convert Units.com, accessed August 26, 2016, www.convertunits.com.

9. Reséndez, *Land So Strange*, 70; Varnum, Álvar Núñez *Cabeza de Vaca*, 50.

10. Rio de las Palmas, known today as Rio Soto la Marina, is 130 miles south of the Rio Grande and 103 miles north of Rio Pánuco, close to the city of Tampico, Mexico. See Varnum, Álvar Núñez *Cabeza de Vaca*, 33–35, 49–52; Reséndez, *Land So Strange*, 70–73; Donald E. Chipman, *Spanish Texas, 1519–1821* (Austin: University of Texas Press, 1992), 79.

11. Reséndez, *Land So Strange*, 64–67, 82–89; Varnum, Álvar Núñez *Cabeza de Vaca*, 52–53.

12. Isla de Malhado (Isle of Misfortune) was the name the Spaniards gave that island. Follet's Island, now a peninsula next to San Luis Pass and across from Galveston Island, is considered the original landing point. See Varnum, Álvar Núñez *Cabeza de Vaca*, 94–106; Reséndez, *Land So Strange*, 128–32, 140–46. Galveston Bay is 745 miles from Apalachicola, Florida. See www.convertunits.com.

13. Chipman and Weddle, "How Historical Myths Are Born," 234; Varnum, Álvar Núñez *Cabeza de Vaca*, 111–14, 123, 129–35, 141–46; Reséndez, *Land So Strange*, 164–67, 173–74, 179–84, 209–10.

14. For the complete story of La Salle and his Texas incursion, see James E. Bruseth and Toni S. Turner, *From a Watery Grave: The Discovery and Excavation of La Salle's Shipwreck*, La Belle (College Station: Texas A&M University Press, 2005); Robert Weddle, *The Wreck of the* Belle, *the Ruin of La Salle* (College Station: Texas A&M University Press, 2001); and William C. Foster, ed., *The La Salle Expedition to Texas: The Journal of Henri Joutel, 1684–1687* (Austin: Texas State Historical Association, 1998).

15. Kenmotsu and Dial, "Native Peoples of the Coastal Prairies"; Welker, "Karankawa Literature"; Bruseth and Turner, *From a Watery Grave*, 25; Weddle, *Wreck of the* Belle, 151–54; Foster, *La Salle Expedition to Texas*, 77–78. A shallop is a small boat for use in shallow water. See "Sailing Vessels of the 18th and 19th Century: Types, Expressions, Parts and Equipment," Bluepete's History of Nova Scotia, accessed September 11, 2016, http://www.blupete.com/Hist/Gloss/Ships.htm.

16. Kenmotsu and Dial, "Native Peoples of the Coastal Prairies"; Foster, *La Salle Expedition to Texas*, 69–79; Weddle, *Wreck of the* Belle, 146–53.

17. The Armada de Barlovento was the Windward Fleet, the Spanish ship patrol within the Gulf. The mines were in New Biscay, part of northern Mexico and New Mexico. See Weddle, *Wreck of the* Belle, 90, 147–48.

18. Weddle, *Wreck of the* Belle, 70–73, 77–79, 90–94, 98–99, 144; Bruseth and Turner, *From a Watery Grave*, 19–20.

19. French Saint-Domingue is present-day Haiti; the Rio Escondido is where the Nueces River is today. The *fleuve* Colbert was the new name of the Mississippi,

Notes to Pages 13–22

in honor of the French foreign minister. See Weddle, *Wreck of the Belle*, 70–73, 78–79, 90–91, 144.

20. Weddle, *Wreck of the* Belle, 139–43, 149, 151–54; Bruseth and Turner, *From a Watery Grave*, 21; Foster, *La Salle Expedition to Texas*, 65.

21. Weddle, *Wreck of the* Belle, 173–74, 179; Bruseth and Turner, *From a Watery Grave*, 3–6.

22. Weddle, *Wreck of the* Belle, 40, 190, 110–11, 159, 185–86; Foster, *La Salle Expedition to Texas*, 102, 130.

23. *Clamcoah* was the French term for the Karankawa Indians. See Foster, *La Salle Expedition to Texas*, 10–12. Confrontations between the French and the Karankawas grew fiercer the longer La Salle remained. See Foster, *La Salle Expedition to Texas*, 88–90, 93–94, 99–100, 105–6, 119–22, 136; Weddle, *Wreck of the* Belle, 188–98, 253.

24. Foster, *La Salle Expedition to Texas*, 148, 151, 197–201, 212; Bruseth and Turner, *From a Watery Grave*, 5; Weddle, *Wreck of the* Belle, 189, 251–53; Robert Weddle, "La Salle's Survivors," *Southwestern Historical Quarterly* 75 (1972): 417–21.

25. Ricklis, *Karankawa Indians of Texas*, 146–47; Chipman, *Spanish Texas*, 79; Weddle, "La Salle's Survivors," 416–19.

26. Ricklis, *Karankawa Indians of Texas*, 42, 146–47; Weddle, "La Salle's Survivors," 416–19.

27. Chipman, *Spanish Texas*, 11; Ricklis, *Karankawa Indians of Texas*, 148, 160.

28. Ricklis, *Karankawa Indians of Texas*, 154–55, 168.

29. Almonte actually used "Carancahuas," a different spelling for the same group. See "Almonte's Statistical Report on Texas," trans. Carlos E. Castañeda, *Southwestern Historical Quarterly* 28 (1925): 195.

30. "Aransaso" was another name for Aransas Bay. The name was originally given to its river, Rio Nuestra Señora de Aranzazu. See William J. Guckian and Ramon N. Garcia, *Soil Survey of San Patricio and Aransas Counties, Texas* (US Department of Agriculture and Texas Agriculture Experiment Station, July 1979), 1; Jerry Thompson, *Cortina: Defending the Mexican Name in Texas* (College Station: Texas A&M University Press, 2007), 11, 255; Mary S. Helm, *Scraps of Early Texas History* (Austin: Eakin Press, 1987), 12; Carol A. Lipscomb, "Karankawa Indians," *Handbook of Texas Online*, accessed August 23, 2016, http://www.tshaonline.org/handbook/online/articles/bmk05; "Almonte's Statistical Report on Texas," 195.

CHAPTER 3

1. Francaviglia, *From Sail to Steam*, 79, 96–97.
2. Robert Weddle, *The French Thorn: Rival Explorers in the Spanish Sea* (College

Station: Texas A&M University Press, 1991), 238; Francaviglia, *From Sail to Steam*, 91–92, 94–95, 105–8.

3. Francaviglia, *From Sail to Steam*, 108–9, 116–18; Frank C. Pierce, *A Brief History of the Lower Rio Grande Valley* (Menosha, WI: George Banta, 1917), 22.

4. Frank Wagner, ed., *Bérenger's Discovery of Aransas Pass: A Translation of Jean Bérenger's French Manuscript by William M. Carroll* (Corpus Christi, TX: Friends of the Corpus Christi Museum, 1983), 21–23; C. Herndon Williams, "First Europeans Mapped Aransas Bay in 1720," *Refugio County Press*, November 21, 2013; Francaviglia, *From Sail to Steam*, 18; Weddle, *French Thorn*, 218–19.

5. Wagner, *Bérenger's Discovery of Aransas Pass*, 27–28; Weddle, *French Thorn*, 34–38, 42, 236–38, 303; Francaviglia, *From Sail to Steam*, 85; C. Herndon Williams, *Texas Gulf Coast Stories* (Charleston, SC: History Press, 2010), 43–44.

6. Pierce, *Brief History of the Lower Rio Grande Valley*, 22; Francaviglia, *From Sail to Steam*, 116, 120–25; George F. Haugh, "Notes and Documents: History of the Texas Navy," *Southwestern Historical Quarterly* 58 (April 1960): 572–73.

7. Ulysses S. Grant, *Personal Memoirs, 1885–1886*, chap. 4, accessed January 25, 2017, http://www.bartleby.com/1011/4.html; Fairfax Downey, *Texas and the War with Mexico* (New York: American Heritage, 1961), 71.

8. Robert Weddle, *Changing Tides: Twilight and Dawn in the Spanish Sea* (College Station: Texas A&M University Press, 1995), 35; James W. Sheire, *Padre Island National Seashore Historic Resource Study* (Washington, DC: US Department of the Interior National Park Service, Office of History and Historic Architecture, August 1971), chap. 1, appendix 1; Francaviglia, *From Sail to Steam*, 72. The final name came from the Balli family, who ranched the island in the early 1800s. Nicolas Balli, primary landowner, was a priest. See C. Pearce Schaudies to the Balli Heirs, March 17, 1967, Renato Ramirez documents; and Tunnell and Judd, *Laguna Madre*, 61.

9. Tradition indicates that Pineda entered the bay on the feast of Corpus Christi and so named it, although there is no evidence of such. See also Francaviglia, *From Sail to Steam*, 72; Art Leatherwood, "Corpus Christi Bay," *Handbook of Texas Online*, accessed September 22, 2016, http://www.tshaonline.org/handbook/online/articles/rrc05; Sheire, *Padre Island National Seashore*, appendix; Weddle, *Changing Tides*, 29; Herb Canales, "¡Viva el Rey Alonso! The Legend of Who Discovered and Named Corpus Christi Bay," *Journal of South Texas* 24 (2011): 63–64; Leroy P. Graf, "Colonizing Projects in Texas South of the Nueces, 1829–1845," *Southwestern Historical Quarterly* 50 (April 1947): 436.

10. Guido Frankl and Ramon N. Garcia, *Soil Survey: Nueces County, Texas*, series 1960, no. 26 (US Department of Agriculture Soil Conservation Service, 1965) 1, 12, 17; Fred Jones, *Flora of the Texas Coastal Bend* (Sinton, TX: Rob and

Bessie Welder Wildlife Foundation, 1982) xiv–xv; W. S. Henry, *Campaign Sketches of the War with Mexico* (New York: Harper Brothers, 1847), 13; Wagner, *Bérenger's Discovery of Aransas Pass*, 23n27; *Corpus Christi Star*, November 28, 1848; G. Joan Holt, ed., with Sharon Herzka and Robert Ricklis, *The State of the Bay: A Report for the Future*, 33, accessed September 27, 2016, www.//cbbep.org/publications/virtuallibrary/StateoftheBayPart2.pdf (site discontinued; printout in author's possession); Murphy Givens and Jim Moloney, *Corpus Christi: A History* (Corpus Christi, TX: Nueces Press, 2011), 11, 15, 20.

11. Frankl and Garcia, *Soil Survey*, 1.

12. The settlement's original 1845 population was estimated to be one hundred. See Grant, *Personal Memoirs*, "Corpus Christi," 2; Darwin Payne, "Camp Life in the Army of Occupation: Corpus Christi, July 1845–March 1846," *Southwestern Historical Quarterly* 73 (January 1970): 326, 335; Ethan Allen Hitchcock, *Fifty Years in Camp and Field: Diary of Major-General Ethan Allen Hitchcock, USA*, ed. W. A. Croffut (New York: G. P. Putnam's Sons, 1909), 195.

13. R. H. Thorton, "Taylor's Trail in Texas," *Southwestern Historical Quarterly* 70 (July 1977): 8; Charles D. Spurlin, *Texas Volunteers in the Mexican War* (Austin, TX: Eakin Press, 1998), 3; Hitchcock, *Fifty Years in Camp and Field*, 197.

14. Murphy Givens, "Was North Beach Called Smugglers' Island?," April 13, 2011; "Henry Kinney: When the Bottom Fell Out in Illinois," October 9, 2013; "Henry Kinney's Early Years in Corpus Christi," October 16, 2013; "Kinney's Lobbying Brought an Army to Corpus Christi," October 23, 2013, all from *Corpus Christi Caller-Times*; Henry, *Campaign Sketches*, 19.

CHAPTER 4

1. There were two treaties signed at Velasco, one public and one secret. Neither was honored, and Mexico refused to recognize Texas independence for the next twelve years. See "Treaties of Velasco," *Handbook of Texas Online*, accessed February 8, 2017, http://www.tshaonline.org/handbook/online/articles/mgt05; A. Ray Stephens and William K. Holmes, *Historical Atlas of Texas* (Norman: University of Oklahoma Press, 1989), 21, 24, 33.

2. Words within parentheses are Marcy's; see William L. Marcy to General Taylor, May 28, July 8, and October 16, 1845, in *General Taylor's Life, Battles and Correspondence* (Philadelphia: T. C. Clarke, 1847), accessed March 16, 2017, https://archive.org/stream/brilliantnationa00phil/brilliantnationa00phil_djvu.t.

3. Marcy to Taylor, May 28, 1845, *General Taylor's Life*; Daniel P. Whiting, *A Soldier's Life: Memoirs of a Veteran of 30 Years of Soldiering, Seminole Warfare in Florida, the Mexican War, Mormon Territory, and the West*, ed. Murphy Givens

(Corpus Christi, TX: Nueces Press, 2011), 65; Grant, *Personal Memoirs*; Charles N. Pede, "Discipline Rather Than Justice: Courts-Martial and the Army of Occupation at Corpus Christi, 1845–1846," *Army History*, Fall 2016, 36.

4. Grant, *Personal Memoirs*; Hitchcock, *Fifty Years in Camp and Field*, 203; "The March of the 2d Dragoons," *Daily Picayune* (New Orleans, LA), September 25, 1845.

5. "Lightering" was a term referring to the shifting of cargo from large boats to smaller craft, like steam packets or skiffs, with access to shoreline harbors. Henry, *Campaign Sketches*, 14; Pede, "Discipline Rather Than Justice," 35; Payne, "Camp Life in the Army," 326; Francaviglia, *From Sail to Steam*, 98, 132, 172; Whiting, *Soldier's Life*, 65.

6. "Corpus Christi Village and the Camp of the Army of Occupation—1845," *New York Herald*, March 14, 1846; Whiting, *Soldier's Life*, 66, 113.

7. Whiting, *Soldier's Life*, 66; Grant, *Personal Memoirs*.

8. George Meade, *The Life and Letters of George Gordon Meade, Major General, United States Army*, vol. 1 (New York: Charles Scribner's Sons, 1913), letter of October 16, 1845; Henry, *Campaign Sketches*, 28.

9. Henry, *Campaign Sketches*, 44; Edward S. Wallace, "General William Jenkins Worth and Texas," *Southwestern Historical Quarterly* 54 (October 1950): 161; Thorton, "Taylor's Trail in Texas"; Murphy Givens, "Kinney's Lobbying Brought an Army to Corpus Christi," *Corpus Christi Caller-Times*, October 23, 2013.

10. Hitchcock, *Fifty Years in Camp and Field*, 203; Abner Doubleday, *My Life in the Old Army*, ed. Joseph Chance (Fort Worth: Texas Christian University Press, 1998), 43–44; Pede, "Discipline Rather Than Justice," 35–39, 40, 48.

11. Payne, "Camp Life in the Army," 329; Grant, *Personal Memoirs*; Pede, "Discipline Rather Than Justice," 38.

12. Payne, "Camp Life in the Army," 328–29.

13. Samuel French, *Two Wars: An Autobiography of General Samuel G. French* (Nashville: Confederate Veteran, 1901), 34; Henry, *Campaign Sketches*, 16; Payne, "Camp Life in the Army," 329; Whiting, *Soldier's Life*, 67.

14. Meade, *Life and Letters*, letter of December 9, 1845; Pede, "Discipline Rather Than Justice," 39; Hitchcock, *Fifty Years in Camp and Field*, 198.

15. The death count is generally agreed to be eight, although E. A. Hitchcock stated ten, and Napoleon Dana listed nine. See Hitchcock, *Fifty Years in Camp and Field*, 200–201; and Napoleon Jackson Tecumseh Dana, *Monterrey Is Ours! The Mexican War Letters of Lieutenant Dana, 1845–1847*, ed. Robert H. Ferrell (Lexington: University Press of Kentucky, 1990), 9–12. The memorial above their graves in Corpus Christi's Old Bayview Cemetery lists eight names; see "Details for Explosion of the Steamship Dayton," *Texas Historic Sites Atlas*, accessed April 2,

2017, https://atlas.thc.state.tx.us/details/5355001519. See also Henry, *Campaign Sketches*, 35–38; Meade, *Life and Letters*, letter of September 14, 1845.

16. Mary C. Gillett, *The Army Medical Department, 1818–1865* (Washington, DC: Government Printing Office, 1987), 99; Meade, *Life and Letters*, letter of December 9, 1845; Dana, *Monterrey Is Ours!*, 17.

17. Meade, *Life and Letters*, letter of October 21, 1845; Henry, *Campaign Sketches*, 45; Dana, *Monterrey Is Ours!*, 19; Whiting, *Soldier's Life*, 67; Grant, *Personal Memoirs*; Hitchcock, *Fifty Years in Camp and Field*, 208.

18. Whiting, *Soldier's Life*, 66; Dana, *Monterrey Is Ours!*, 7, 16–17.

19. "Catarrhal fever" was the term used at the time for respiratory diseases like the common cold, influenza, and forms of pneumonia. See The Free Dictionary, accessed January 27, 2017, http://medical-dictionary.thefreedictionary.com/catarrhal+fever; also see Gillett, *Army Medical Department*, 99; Payne, "Camp Life in the Army," 331, 333; Pede, "Discipline Rather Than Justice," 38–39; Meade, *Life and Letters*, letter of October 21, 1845; Doubleday, *My Life in the Old Army*, 45–46; Henry, *Campaign Sketches*, 32; Whiting, *Soldier's Life*, 67.

20. Payne, "Camp Life in the Army," 333; Grant, *Personal Memoirs*.

21. Henry, *Campaign Sketches*, 52–53.

CHAPTER 5

1. Marcy to Taylor, October 16, 1845, *General Taylor's Life*; Meade, *Life and Letters*, letters of October 9, October 10, November 12, 1845, January 26, 1846.

2. Gordon A. Atwater, "Navigation," *Collier's Encyclopedia* (USA: Crowell-Collier Education Corp., 1969), 228–29; "Navigation in the 18th Century," Penobscot Marine Museum, accessed November 12, 2016, https://penobscotmarinemuseum.org/pbho-1/history-of-navigation/navigation-18th-century.

3. Meade, *Life and Letters*, letters of February 18, February 24, 1846.

4. Henry, *Campaign Sketches*, 34, 38; Francaviglia, *From Sail to Steam*, 132; Whiting, *Soldier's Life*, 65; Hitchcock, *Fifty Years in Camp and Field*, 193–94, 200; Grant, *Personal Memoirs*.

5. *Aranzas Bay as Surveyed by Captain Monroe of the Amos Wright*, 1833, chart courtesy of Port Aransas Preservation and Historical Association, hereafter cited as PAPHA or Port Aransas Museum; *Mustang Island Map of 1846*, chart courtesy of the University of Texas Marine Science Institute.

6. Grant, *Personal Memoirs*; Hitchcock, *Fifty Years in Camp and Field*, 193–94; Meade, *Life and Letters*, letter of September 6, 1845.

7. Grant, *Personal Memoirs*; Hitchcock, *Fifty Years in Camp and Field*, 194–97; Henry, *Campaign Sketches*, 30.

8. Meade, *Life and Letters*, letter of January 21, 1846; M. O. Crimmins, ed., "Notes and Documents: W. G. Freeman's Report on the Eighth Military Department," *Southwestern Historical Quarterly* 50 (1947): 352; "Coastal Texas II," Texas Historical Sites, accessed April 3, 2017, http://www.northamericanforts.com/West/tx-coast2.html; "Camp Corpus Christi," *Handbook of Texas Online*, accessed April 3, 2017, http://www.tshaonline.org/handbook/online/articles/qbc10.

9. *Aranzas Bay as Surveyed by Captain Monroe*; US Coast Survey-1851, Sketch 1, No. 3, Aransas Pass, Coast of Texas Reconnaissance; US Coast Survey-1853, Sketch 1, no. 5, Reconnaissance of Aransas Pass Texas, all courtesy of PAPHA; Tom W. Stewart, "History of the Aransas Pass Jetties" (presentation at the Texas Branch of the American Society of Civil Engineers, Corpus Christi, TX, March 22, 2013).

10. "Negro Citizens," October 3, 1857; "The Northern Democracy," October 17, 1857; "The Negro," March 27, 1848, all from the *Nueces Valley*; "Unprotected State of the Frontier—Want of Cavalry at Laredo, Fort Ewell and Rio Grande City"; "Guerrillas Attack Brownsville," October 22, 1859; "The Frontier—Our Situation and Our Wants," November 19, 1859; "A New Commander of the Department of Texas," March 3, 1860; "Important Orders from the War Department," March 10, 1860, all from the *Ranchero* (Corpus Christi, TX).

11. "Abolition Incendiaries in Texas," August 4, 1860; "Letter from Dallas," August 11, 1860; "More Abolition Outrages," September 1, 1860; "Letter from Captain Lovenskiold," February 18, 1860; "Corpus Christi, January 22," January 26, 1861, all from the *Ranchero* (Corpus Christi, TX).

12. Williams, *Texas Gulf Coast Stories*, 71; Linda Wolff, *Indianola and Matagorda Island: 1837–1887* (Austin: Eakin Press, 1999), 38; John Guthrie Ford, ed., *The Mercer Logs: Pioneer Times on Mustang Island, Texas*, no. 35 (Port Aransas, TX: Port Aransas Preservation and Historical Association, 2012), 12; Norman C. Delaney, *The Maltby Brothers' Civil War* (College Station: Texas A&M University Press: 2013), 57, 69–70.

13. Williams, *Texas Gulf Coast Stories*, 71; *US Coast Survey-1853, Sketch 1, no. 5, Reconnaissance of Aransas Pass Texas*; "Corpus Christi, Jan. 23," January 26, 1861; "God Bless the Poor Sailor," November 6, 1859; "Aransas Bar: Interesting to the Public," December 3, 1859; "The Aransas Road Company, January 28, 1860; untitled, May 5, 1860, all from the *Ranchero* (Corpus Christi, TX); "History of Dredging," Start Dredging.com, accessed February 27, 2018, https://www.startdredging.com; Holt, *State of the Bay*, 11–14, 27, 34.

14. Ford, *Mercer Logs*, 12, 14, 23; Jones, *Flora of the Texas Coastal Bend*, xvii–xx; Tunnell and Judd, *Laguna Madre*, 69.

15. According to one of his officers, Henry Sibley's ultimate goal, following the Confederate seizure of New Mexico territory, "was the conquest of California

and the annexation of northern Mexico." See Martin Hardwick Hall, *Sibley's New Mexico Campaign* (Albuquerque: University of New Mexico Press, 1960), 23.

16. Maria von Blücher, *Maria von Blücher's Corpus Christi: Letters from the South Texas Frontier, 1849–1879*, ed. Bruce S. Cheeseman (College Station: Texas A&M University Press, 2002), 129; Eugenia Reynolds Briscoe, *City by the Sea: A History of Corpus Christi* (New York: Vantage Press, 1985), 199–200; Richard N. Current, "U. S. Civil War," *Collier's Encyclopedia*, vol. 6, ed. William D. Halsey (USA: Crowell-Collier Educational Corp., 1969), 527; Mark Carnes and John Garraty, *Mapping America's Past* (New York: Henry Holt, 1996), 114; John Keegan, *The American Civil War: A Military History* (New York: Alfred A. Knopf, 2009), 137; "Still Another Confederate Disaster," *San Antonio Herald*, March 1, 1862; "Our Situation," *Galveston Tri-Weekly News*, May 3, 1862.

17. Delaney, *Maltby Brothers' Civil War*, 63, 69–70; J. W. Kittredge to D. G. Farragut, October 15, 1862, *Official Records of the Union and Confederate Navies in the War of the Rebellion*, series 1, vol. 19 (Washington, DC: Government Printing Office, 1905), accessed April 6, 2017, https://texashistory.unt.edu/ark:/67531/metapth192854/?q=Sabine%20Pass, hereafter cited as *O.R.*

CHAPTER 6

1. Delaney, *Maltby Brothers' Civil War*, 63, 70; Norman Delaney, "Whiskey and the Battle of Corpus Christi," The Texas Story Project, Bullock Texas State History Museum, accessed April 7, 2017, https://www.thestoryoftexas.com/discover/texas-story-project/whiskey-shells-corpus-christi; Norman Delaney, "154 Years Ago: A City under Siege," *Corpus Christi Caller-Times Forum*, August 16, 2016, accessed April 8, 2017, http://archive.caller.com/opinion/forums/154-years-ago-a-city-under-seige-3a1f023e-4be8-1759-e-53-0100007fd885-390363691.html.

2. Delaney, *Maltby Brothers' Civil War*, 72; Holland McComb, "The Mexicans Called It 'The Confederate War,' but It Made Matamoros the Cotton Trading Capital of the World," *Corpus Christi Caller-Times*, January 18, 1959; Givens and Moloney, *Corpus Christi*, 51–53; Briscoe, *City by the Sea*, 204.

3. Kittredge to Farragut, October 15, 1862, *O.R.*; Ford, *Mercer Logs*, 16; Delaney, *Maltby Brothers' Civil War*, 63–66; Murphy Givens, "Village on St. Joseph's Was a Victim of War," *Corpus Christi Caller-Times*, December 6, 2006.

4. Delaney, *Maltby Brothers' Civil War*, 66, 70–71; Kittredge to Farragut, October 15, 1862, *O.R.*

5. *Weekly Texas State Gazette* (Austin), March 22, 1862; Colonel H. E. McCullough to Major Samuel Boyer Davis, March 3, 1862, 702; March 25, 1862, 704; March 31, 1862, 705, all from *The War of the Rebellion: A Compilation of the*

Official Records of the Union and Confederate Armies, series 1, vol. 9 (published by an act of Congress, June 16, 1880); Robert Patterson Felgar, "Texas in the War for Southern Independence" (master's thesis, University of Texas, 1935), 84, 206; "Call on Texas for 15,000 More Troops," *San Antonio Herald*, March 1, 1862.

6. "Summary Justice," *San Antonio Herald*, May 31, 1862; Norman Delaney, "Hanging Dedication" (speech delivered at the Texas State Historical Marker dedication, Corpus Christi, TX, May 4, 2017); Norman Delaney, "Two Civil War Hangings in Corpus Christi," *Nueces County Historical Commission Bulletin* 6 (March 2016): 33; Delaney, *Maltby Brothers' Civil War*, 67; Carl Moneyhon, *Edmund J. Davis of Texas: Civil War Leader, Reconstruction Governor* (Fort Worth: Texas Christian University Press, 2010), 43–45.

7. Kittredge to Farragut, October 15, 1862, *O.R.*; Delaney, *Maltby Brothers' Civil War*, 66; Delaney, "154 Years Ago"; "Field Artillery in the Civil War," cwartillery.com, accessed May 13, 2017, https://cwartillery.com/FA/FA.html; "An Alchemists Glossary of Terms, Definitions, Formulas & Concoctions - Part 2," The Third Millennium, accessed May 27, 2017, http://www.3rd1000.com/alchemy/alchemy terms2.htm#N; "Facts about Sulfur," Live Science, accessed May 15, 2017, http://www.livescience.com/28939-sulfur.html.

8. Delaney, *Maltby Brothers' Civil War*, 69, 71; "The Battle of Corpus Christi," The American Civil War, accessed April 8, 2017, https://www.mycivilwar.com/battles/620816.html; Delaney, "154 Years Ago"; Briscoe, *City by the Sea*, 208–9, 211.

9. Delaney, *Maltby Brothers' Civil War*, 73; Kittredge to Farragut, August 12, August 18, October 15, 1862, *O.R.*; Briscoe, *City by the Sea*, 209–10.

10. von Blücher, *Maria von Blücher's Corpus Christi*, 130; Kittredge to Welles, August 13, 1862, *O.R.*; Briscoe, *City by the Sea*, 210–11; Givens and Moloney, *Corpus Christi*, 54–55.

11. Kittredge to Welles, August 13, 1862, *O.R.*; Givens and Moloney, *Corpus Christi*, 54–55; Briscoe, *City by the Sea*, 211; "Battle of Corpus Christi."

12. von Blücher, *Maria von Blücher's Corpus Christi*, 130; Givens and Moloney, *Corpus Christi*, 55; Briscoe, *City by the Sea*, 211.

13. Kittredge to Welles, August 18, 1862, *O.R.*

14. Delaney, *Maltby Brothers' Civil War*, 74; von Blücher, *Maria von Blücher's Corpus Christi*, 131; Delaney, "154 Years Ago"; "Battle of Corpus Christi"; Kittredge to Welles, August 16, 1862, *O.R.*

15. Over the years, rumors spread that many nonexploding shells contained whiskey stolen from Kittredge by his crewmen and secreted within the hollowed-out projectiles. See Delaney, "Whiskey and the Battle." The story appears to be apocryphal. The original account, although quoting Kittredge, was not published until three years after his death; Maria von Blücher reported no such "duds"

in her firsthand description of unexploded shells; and ammunition used during this period was notoriously unreliable, with up to 50 percent failures in some conflicts. See Delaney, "Whiskey and the Battle"; von Blücher, *Maria von Blücher's Corpus Christi*, 130; and "Artillery," US Army Ordnance Corps, accessed May 13, 2017, http://www.goordnance.army.mil/history/Staff%20Ride/STAND%203%20 ARTILLERY%20AND%20SMALL%20ARMS/ARTILLERY%20IN%20 THE%20CIVIL%20WAR.pdf.

16. Kittredge to Welles, August 16, 1862, *O.R.*; von Blücher, *Maria von Blücher's Corpus Christi*, 130; "Battle of Corpus Christi"; Briscoe, *City by the Sea*, 211–12; Kittredge to Welles, August 17, August 18, 1862, *O.R.*; Delaney, *Maltby Brothers' Civil War*, 74.

17. Kittredge to Welles, August 18, 1862, *O.R.*

18. Kittredge to Welles, August 18, 1862, *O.R.*; von Blücher, *Maria von Blücher's Corpus Christi*, 131; "Battle of Corpus Christi"; Delaney, *Maltby Brothers' Civil War*, 75; "Civil War Weapons," History.net, accessed May 13, 2017, https://www.history net.com/civil-war-weapons.

19. von Blücher, *Maria von Blücher's Corpus Christi*, 131; Kittredge to Welles, August 18, 1862, *O.R.*

20. Delaney, *Maltby Brothers' Civil War*, 66; Major E. F. Gray to Brigadier-General H. P. Bee, September 16, 1862; Brigadier-General H. P. Bee to Brigadier-General P. O. Hebert, September 24, 1862, *O.R.*; von Blücher, *Maria von Blücher's Corpus Christi*, 131.

21. Brian Swartz, "Re-enlisting in Texas Could Get a Soldier Home to Maine," Maine at War, accessed April 3, 2017, https://www.maineatwar.bangordailynews .com/2014/01/23 (site discontinued; transcription in author's possession); Delaney, *Maltby Brothers' Civil War*, 84–85; Henry A. Shorey, *The Story of the Maine Fifteenth, Being a Brief Narrative of the More Important Events in the History of the Fifteenth Maine Regiment* (Bridgton, ME: Press of the *Bridgton News*, 1860), 53.

22. "Padre Island National Seashore: The Civil War," US National Park Service, accessed April 8, 2017, https://www.nps.gov/pais/learn/historyculture/civil-war .htm; Delaney, *Maltby Brothers' Civil War*, 78–80.

23. Shorey, *Story of the Maine Fifteenth*, 53, 54; J. Guthrie Ford, "Fort Semmes," *Handbook of Texas Online*, accessed April 3, 2017, http://www.tshaonline .org/handbook/online/articles/qcf29; Delaney, *Maltby Brothers' Civil War*, 84–85; Dana, *Monterrey Is Ours!*, 7.

24. Shorey, *Story of the Maine Fifteenth*, 54–55.

25. Shorey, 57.

26. Shorey, 58; Swartz, "Re-enlisting in Texas."

27. Delaney, *Maltby Brothers' Civil War*, 86.

28. Delaney, 79–81, 86.

29. Delaney, 81, 87.

30. "Quaker Guns" were logs with their muzzles painted black and positioned to deceive the enemy. See Delaney, *Maltby Brothers' Civil War*, 86.

31. The small rebel encampment at the tip of Padre Island had been abandoned by this time, and what remained at Shell Bank Island seems to have been as well. See Delaney, *Maltby Brothers' Civil War*, 85–87.

32. Shorey, *Story of the Maine Fifteenth*, 59–60; Delaney, *Maltby Brothers' Civil War*, 88.

33. Shorey, *Story of the Maine Fifteenth*, 59–60; Ford, "Fort Semmes"; *1851 Port Aransas Confederate*, chart courtesy of PAPHA; Delaney, *Maltby Brothers' Civil War*, 80.

34. Delaney, *Maltby Brothers' Civil War*, 88; Shorey, *Story of the Maine Fifteenth*, 60, 61–63.

35. Shorey, *Story of the Maine Fifteenth*, 62.

36. Other sources indicate the depth was one hundred yards. See Swartz, "Re-enlisting in Texas"; Shorey, *Story of the Maine Fifteenth*, 63.

37. Shorey, *Story of the Maine Fifteenth*, 65; Swartz, "Re-enlisting in Texas"; Ford, "Fort Semmes."

38. Norman C. Delaney, "Civil War Letters Reveal Camp Life on Island," *Corpus Christi Caller-Times*, July 24, 2006; Robin Borglum Kennedy, *Mary's Story: Mary Borglum's Story from the Mountains of Anatolia to the Mountains of South Dakota* (North Charleston, SC: CreateSpace, 2013), 25–26.

CHAPTER 7

1. T. G. Barragy, *Gathering Texas Gold: Frank Dobie and the Men Who Saved the Longhorns* (Corpus Christi, TX: Cayo del Grullo Press, 2003), 22–25, 29–39; Donald E. Worcester, "Longhorn Cattle," *Handbook of Texas Online*, accessed August 21, 2017, https://www.tshaonline.org/handbook/entries/longhorn-cattle; "Hereford," The Cattle Site, accessed August 20, 2017, www.thecattlesite.com/breeds/beef/14/hereford.

2. Jimmy M. Skaggs, "Cattle Trailing," *Handbook of Texas Online*, accessed August 20, 2017, http://www.tshaonline.org/handbook/online/articles/ayc01.

3. Tom Lea, *The King Ranch*, vol. 1 (Boston: Little, Brown, 1957), 2–8; Bruce S. Cheeseman, "Richard King," *Handbook of Texas Online*, accessed January 18, 2008, https://www.tshaonline.org/handbook/entries/king-richard.

4. Lea, *King Ranch*, 21, 26–27, 35–36, 54, 57–59, 70–76.

5. Lea, 100, 156–57, 186; Cheeseman, "Richard King"; "Steamer 'James Hall'

Ashore," January 19, 1861; "A Card," January 26, 1861, both from the *Ranchero* (Corpus Christi, TX).

6. The final sale of the riverboat business occurred in 1874. See Bob Kinnan, "Echoes in the Dust: Captain Richard King and Rancho de Santa Gertrudis, 1865–1885" (presentation at the East Texas Historical Association, Galveston, TX, October 13, 2017). See also Lea, *King Ranch*, 251, 255, 304; Cheeseman, "Richard King"; Tunnell and Judd, *Laguna Madre*, 66, 67.

7. By the end of his life, King owned over five hundred thousand acres of land. As well as cattle, livestock driven to market included horses and mules. See Kinnan, "Echoes in the Dust"; Lea, *King Ranch*, 301–3, 329; Cheeseman, "Richard King"; Jay Nixon, *Stewards of a Vision: A History of King Ranch* (Hong Kong: King Ranch, 1986), 14; *King Ranch: 100 Years of Ranching, 1853–1953* (Corpus Christi, TX: *Corpus Christi Caller-Times*, 1953), 17.

8. That amount is the equivalent of $10,587,921.30 in 2016 currency. See inflation calculator, Official Inflation Data, Alioth Finance, accessed September 8, 2017, http://www.in2013dollars.com/1875-dollars-in-2016?amount=500000; Kinnan, "Echoes in the Dust"; Lea, *King Ranch*, 298–99, 305, 319–21; Cheeseman, "Richard King."

9. Lea, *King Ranch*, 300; Martin Donell Kohout, "Duval County," *Handbook of Texas Online*, accessed September 8, 2017, http://www.tshaonline.org/handbook/online/articles/hcd11; Givens and Moloney, *Corpus Christi*, 119–20.

10. "My Annual Stroll," *Nueces Valley*, September 25, 1858; "Early North Beach Packer Kept Hides, Tallow, Discarded Meat," *Corpus Christi Caller-Times*, 1952 [only the year was given]; Givens and Moloney, *Corpus Christi*, 121–23; Briscoe, *City by the Sea*, 278; Lea, *King Ranch*, 304.

11. Lea, *King Ranch*, 132, 303, 381, 411; Jones, *Flora of the Texas Coastal Bend*, xiv–xx.

12. The purchase later appeared to be of dubious authenticity and the source for King's insistence on legally scrutinized transactions from that time on. By 1905, only a certain amount of the Padre Island tract was accepted as King Ranch property. See Cheeseman, "Richard King"; and Lea, *King Ranch*, 438n31.

13. The amount King would have put out, in today's value, would have been $5,406.60. See inflation calculator, Official Inflation Data, Alioth Finance, accessed September 17, 2017, http://www.in2013dollars.com/1854-dollars-in-2016?amount=200; Lea, *King Ranch*, 132, 417.

14. Greg Smith, interview by author, March 5, 2018, Corpus Christi, TX; Greg Smith, "Ranching on the Islands" (presentation at the South Texas Historical Association, Port Aransas, TX, November 2, 2013); Greg Smith, "The Dunn Family in Nueces County and Ranching on Padre Island" (presentation at the Nueces

County Historical Society, Corpus Christi, TX, April 1, 2014); "Padre Island National Seashore: Dunn Ranch," National Park Service, accessed September 18, 2017, https://www.nps.gov/pais/learn/historyculture/dunn-ranch.htm; Murphy Givens, *Great Tales from the History of South Texas* (Corpus Christi, TX: Nueces Press, 2012), 305, 309.

15. Encinal Peninsula is more commonly known as Flour Bluff. Smith, interview; Smith, "Dunn Family"; Givens, *Great Tales*, 307; Jack E. Davis, *The Gulf: The Making of an American Sea* (New York: Liveright, 2017), 320.

16. "Padre Island National Seashore: Dunn Ranch."

17. Givens and Moloney, *Corpus Christi*, 121–23; Alice M. Shukalo, "Rockport, TX," *Handbook of Texas Online*, accessed July 24, 2017, http://www.tshaonline.org/handbook/online/articles/hgr05.

18. Briscoe, *City by the Sea*, 278.

19. Givens and Moloney, *Corpus Christi*, 122–23.

20. Live cattle were still being shipped from the bay. See *Improvement of Aransas Pass and Bay up to Rockport and Corpus Christi*, report of Major Ernst, Chief of Engineers, US Army (June 1887), 1432. In context, this selection appears to come from appendix T-7 in *Chief of Engineers United States Army, to the Secretary of War for the Year 1888, in Four Parts*, Part 2 (Washington, DC: Government Printing Office, 1888), courtesy of PAPHA; *Centennial History of Corpus Christi* (Corpus Christi, TX: *Corpus Christi Caller-Times*, 1952), 88.

21. James P. Baughman, *Charles Morgan and the Development of Southern Transportation* (Nashville: Vanderbilt University Press, 1968), 129; *Centennial History of Corpus Christi*, 88; Briscoe, *City by the Sea*, 278.

22. "Marine Intelligence," *Daily Herald* (San Antonio, TX), February 2, 1866; "From Galveston: Marine Intelligence," *Daily Herald* (San Antonio, TX), February 15, 1855; "Morgan Line," *Flake's Daily Bulletin* (Galveston, TX), September 1, 1866; C. Herndon Williams, "The Star of St. Mary's of Aransas Never Went Out," *Baysider*, August 13, 2016; Baughman, *Charles Morgan*, 23, 50–51, 126–27, 129–30; Francaviglia, *From Sail to Steam*, 223, 230.

CHAPTER 8

1. Kellie Crnkovich, comp., "Robert Ainsworth Mercer," 1, accessed October 10, 2008, http://www.rootsweb.ancestry.com/~txaransa/mercer.htm; "Lancashire - County Maps," MapStop Global Mapping, accessed September 30, 2017, www.mapstop.co.uk/product10439_Lancashire-County-Maps.aspx; "Robert Mercer," Lancashire, England, Church of England Marriages and Banns, 1754–1936, Ancestry.com, 2012.

2. P. William Filby, ed., *Passenger and Immigration Lists Index, 1500s–1900s* (Farmington Hill, MI: Gale Research, 2012); *Fifth Census of the United States, 1830*; *Sixth Census of the United States, 1840*, both from Records of the Bureau of the Census; "Historic New Albany," accessed September 29, 2017, www.historicnewalbany.com; "Overview of the Lafayette Township, Floyd County, Indiana (Township)," Statistical Atlas, accessed September 29, 2017, https://statisticalatlas.com/county-subdivisions/Indiana (site discontinued; transcript in author's possession).

3. *Seventh Census of the United States, 1850*, Records of the Bureau of the Census; Crnkovich, "Robert Ainsworth Mercer," 3–6.

4. 1867 Voter Registration Lists, microfilm, Texas State Library and Archives Commission, Austin; *Ninth Census of the United States, 1870*, Records of the Bureau of the Census; Crnkovich, "Robert Ainsworth Mercer," 1.

5. Crnkovich, "Robert Ainsworth Mercer," 1.

6. Crnkovich, 2, 4.

7. The word "bar" referred to the sand buildups that prevented easy access across barrier isle passes. "Bar pilots" has been shortened to "pilots." Paul G. Kirchner and Clayton L. Diamond, "Unique Institutions, Indispensable Cogs, and Hoary Figures: Understanding Pilotage Regulation in the United States," *University of San Francisco School of Law Maritime Law Journal* 23, no. 1 (2010–2011), accessed August 4, 2017, http://www.americanpilots.org/document_center/Activities/Unique_Institutions__Indispensable_Cogs__and_Hoary_Figures_Understanding_Pilotage_Regulation_in_the_United_States.pdf; Stewart, "History of the Aransas Pass Jetties."

8. *Bisso v. Inland Waterways Corporation*, 50 US (1955), FindLaw, accessed September 30, 2017, http://caselaw.findlaw.com/us-supreme-court/349/85.html.

9. *Improvement of Aransas Pass*, 1431; Phyllis Coffee, "Logs Reveal Texas Gulf Coast History, 1866–1900," *Southwestern Historical Quarterly* 62 (1959): 229; Stewart, "History of the Aransas Pass Jetties"; *Coast Chart No. 210–1890: 1868–1875 Surveys*; *U.S. Coast Survey: Corpus Christi Pass, Texas*, surveyed by H. Anderson, 1869, charts, University of Texas Marine Science Institute, Port Aransas.

10. Ford, *Mercer Logs*, 21, 23, 25, 43, 63.

11. Ford, *Mercer Logs*, 51, 53; Crnkovich, "Robert Ainsworth Mercer," 9; Phyllis Coffee, "After 90 Years, 5 Old Diaries Make News Again," *Corpus Christi Times*, August 5, 1957; Stewart, "History of the Aransas Pass Jetties."

12. Baughman, *Charles Morgan*, 185; Briscoe, *City by the Sea*, 278; Stewart, "History of the Aransas Pass Jetties"; Givens and Moloney, *Corpus Christi*, 123.

13. Ford, *Mercer Logs*, 53, 54.

14. Ford, 52, 64.

15. *United States Coast Pilot: Atlantic Coast*, part 7, *Gulf of Mexico from Key West to the Rio Grande* (Washington, DC: US Treasury Department, US Coast and

Geodetic Survey Office, 1896), 121; Ford, *Mercer Logs*, 52, 86; Stewart, "History of the Aransas Pass Jetties."

16. The equivalent of $28 in 2016 currency is $564.69; see inflation calculator, Official Inflation Data, Alioth Finance, http://www.in2013dollars.com/1874-dollars-in-2016?amount=28. A draft was the distance between the waterline of the vessel and the bottom of the hull. The draft or draught indicated how much water had to be in the pass to allow the pilot to cross it safely. "Steamboats: Texas; Morgan Line U. S. Mail Steamers"; "Passengers and Mail Routes," both from *New Orleans Daily Democrat*, October 30, 1877; *Improvement of Aransas Pass*; Jack B. Irion and David A. Ball, "The *New York* and the *Josephine*: Two Steamships of the Charles Morgan Line," Bureau of Ocean Energy Management, accessed October 7, 2017, www.boem.gov/uploadedFiles/BOEM/Environmentsal/Stewardship/Archeology/Irion-and-Bell.pdf (site discontinued; transcript in author's possession); "19th Century Steamships," Bureau of Energy Management, accessed October 7, 2017, https://www.boem.gov/environment/19th-century-steamships; David G. McComb, *Galveston: A History* (Austin: University of Texas Press, 1986), 57; *United States Coast Pilot*, 105.

17. Francaviglia, *From Sail to Steam*, 140; Givens and Moloney, *Corpus Christi*, 123.

18. *Constitution of the State of Texas, Adopted by the Constitutional Convention Convened at Austin, September 6, 1875, and Ratified by the People, February 15, 1876*, in Henry F. Triplett and Ferdinand A. Hauslein, *Civics: Texas and Federal* (Houston: Rein and Sons, 1912), 200; Kirchner and Diamond, "Unique Institutions"; John T. Thompson, "Governmental Responses to the Challenges of Water Resources in Texas," *Southwestern Historical Quarterly* 70 (1966): 45; Ford, *Mercer Logs*, 37–38, 45.

19. Ford, *Mercer Logs*, 52, 54, 65, 66–67, 71, 88–89.

20. Stewart, "History of the Aransas Pass Jetties"; *US Coast Survey-1853, Sketch 1, no. 5, Reconnaissance of Aransas Pass Texas*, chart courtesy of PAPHA; Ford, *Mercer Logs*, 55, 56.

21. James Poskett, "Edward Massey," *Longitude Essays*, Cambridge Digital Library, accessed August 27, 2018, https://cudl.lib.cam.ac.uk/view/ES-LON-00028/1; Boban Docevski, "Depth Sounding Techniques That Preceded the Modern Day SONAR Technology," *Vintage News*, February 23, 2017, accessed August 27, 2018, https://www.thevintagenews.com/2017/02/23/depth-sounding-techniques-that-preceded-the-modern-day-sonar-technology/.

22. Stewart, "History of the Aransas Pass Jetties"; "Beacons and Buoys Needed," *Corpus Christi Caller*, November 28, 1886; Francaviglia, *From Sail to Steam*, 238–39; *United States Coast Pilot*, 106; Coffee, "Logs Reveal Texas Gulf Coast History," 230.

23. John J. Mayo, *The British Code List for 1874 for the Use of Ships at Sea, and for Signal Stations* (London: Sir William Mitchell, 1874), a, a2, iii, courtesy of PAPHA; "Pilot Flags," Flags of the World, accessed October 21, 2019, http://www.crwflags.com/fotw/flags/xf-pilt.html.

24. Mayo, *British Code List*, a, ix; "Pilot Flags."

25. Mayo, iv, ix, xii.

26. The cost of the *British Code List* was five shillings, roughly equivalent to $28 in 2017 dollars. See X-Rates, accessed October 21, 2017, http://www.x-rates.com/calculator/?from=GBP&to=USD&amount=21.

27. C. C. Heath and William R. Roberts, "Code of Signals" flyer, August 1, 1874, courtesy of PAPHA.

28. The *Aransas* needed clearance of ten feet above the bar when it entered the pass. See "Commercial Statistics of Aransas Pass, Texas, for the Fiscal Year Ending June 30, 1888," in *Chief of Engineers United States Army, to the Secretary of War for the Year 1888, in Four Parts*, part 2 (Washington, DC: Government Printing Office, 1888), 1311, hereafter cited as *CofE*.

29. Heath and Roberts, "Code of Signals."

30. "The Storm Victims," *Corpus Christi Caller*, August 29, 1886; Francaviglia, *From Sail to Steam*, 184, 238.

31. McComb, *Galveston*, 57, 58–61, 118; Francis E. Sargent and Robert R. Bottin Jr., *Case Histories of Corps Breakwater and Jetty Structures* (Washington, DC: Department of the Army, US Corps of Engineers, 1989), 4, 6, 17.

32. The line of breakers formed by these shoals was termed "riffles." See Ford, *Mercer Logs*, 52.

33. Ford, *Mercer Logs*, 51.

34. Ford, *Mercer Logs*, 57; Francaviglia, *From Sail to Steam*, 227–28; Coffee, "Logs Reveal Texas Gulf Coast History," 230.

35. Coffee, "After 90 Years, 5 Old Diaries Make News Again," *Corpus Christi Times*, August 5, 1957; Ford, *Mercer Logs*, 51, 57–58.

36. Phyllis Coffee, "Mercer Logs: The Story of an Ill-Fated Vessel," *Corpus Christi Times*, August 6, 1957; Coffee, "Logs Reveal Texas Gulf Coast History," 229.

37. National Weather Service Heritage, "NWS Timeline," accessed December 9, 2002, https://vlab.noaa.gov/web/nws-heritage/explore-nws-history#event-debut-of-the-daily-weather-map.

38. Ford, *Mercer Logs*, 63–65.

39. Ford, *Mercer Logs*, 30, 65–66; Coffee, "After 90 Years."

40. Ford, *Mercer Logs*, 45, 48.

41. Ford, 67.

42. Ford, *Mercer Logs*, 70; "More Particulars of the Blow: A Vivid Description

of the Ruins at Indianola," August 22, 1886; "A Terrific Rain Storm," August 29, 1886, both from *Corpus Christi Caller*.

43. "Death and Destruction at Sabine," *Corpus Christi Caller*, October 17, 1886.

44. Quoted in "The Texas Storms," *Corpus Christi Caller*, October 24, 1886.

CHAPTER 9

1. Ford, *Mercer Logs*, 43.

2. "Spat" is the term for the third stage of oyster development. See "Oyster Life Cycle," University of Maryland Center for Environmental Science, accessed November 26, 2018, https://hatchery.hpl.umces.edu/oyster-life-cycle/; also "Fish and Oysters," *Daily Herald* (San Antonio, TX), March 8, 1866; Venable et al., *Historical Review of the Nueces Estuary*, 2; Tunnell et al., *Encyclopedia of Texas Seashells*, 41–42.

3. "Proceedings of the City Council," March 14, 1886; "Proposals for Shells," March 28, 1886, both from *Flake's Daily Bulletin* (Galveston, TX), 5; "Street Improvements," *Corpus Christi Caller*, October 10, 1886; Tunnell et al., *Encyclopedia of Texas Seashells*, 24–25; Venable et al., *Historical Review of the Nueces Estuary*, 5.

4. Holt, *State of the Bay*, 33; Tunnell et al., *Encyclopedia of Texas Seashells*, 14–17, 41–43; "History of Dredging."

5. Jeremiah Frederick Kratz, "A History of the Texas Shrimp Industry" (master's thesis, University of Texas, 1963), 4, 6, 8–10.

6. Paul A. Montagna, Scott Holt, Christine Ritter, Sharon Herzka, and Kenneth H. Dunton, *Characterization of Anthropogenic and Natural Disturbance on Vegetated and Unvegetated Bay Bottom Habitats in the Corpus Christi Bay National Estuary Program Study Area* (Port Aransas: University of Texas at Austin Marine Science Institute, May 1998), 6, accessed September 27, 2016, www.//cbbep.org/publications/virtuallibrary/cc25a.pdf; Kratz, "History of the Texas Shrimp Industry," 6, 10–11, 15, 19.

7. The turtles got their name from the color of the fat under their shells. See "Information about Sea Turtles: Green Sea Turtles," Sea Turtle Conservancy, accessed December 18, 2017, https://conserveturtles.org; "Sea Turtles: Adaptations," SeaWorld, accessed December 18, 2017, https://seaworld.org/animals/all-about/sea-turtles/adaptations/.

8. "Early North Beach Packer Kept Hides, Tallow, Discarded Beef," *Corpus Christi Caller-Times*, 1952; Givens and Moloney, *Corpus Christi*, 121–23; Simon Freese and Deborah Lightfood Sizemore, *A Century in the Works: Freese and Nichols Consulting Engineers* (College Station: Texas A&M University Press, 1994), 5; Jones, *Flora of the Texas Coastal Bend*, xx; Robin W. Doughty, "Sea Turtles in

Texas: A Forgotten Commerce," *Southwestern Historical Quarterly* 88 (July 1984): 53–54; "Information about Sea Turtles."

9. Givens and Moloney, *Corpus Christi*, 123; Cheeseman, "Richard King"; Doughty, "Sea Turtles in Texas," 50–53.

10. Doughty, "Sea Turtles in Texas," 47, 51, 53–55, 61, 62.

11. Doughty, "Sea Turtles in Texas," 66. In 2016, the US Endangered Species Act listed green sea turtles as still likely to be in danger of extinction. See "Information about Sea Turtles."

12. Holt, *State of the Bay*, 12; Doughty, "Sea Turtles in Texas," 47–48.

13. Coffee, "Logs Reveal Texas Gulf Coast History," 230–31; Doughty, "Sea Turtles in Texas," 48, 61; Kratz, "History of the Texas Shrimp Industry," 9, 15.

14. In 2016 currency, this would be equivalent to $2,853,166.16. See inflation calculator, Official Inflation Data, Alioth Finance, accessed December 28, 2017, http://www.in2013dollars.com/1880-dollars-in-2016?amount=128000; Kratz, "History of the Texas Shrimp Industry," 23.

CHAPTER 10

1. McComb, *Galveston*, 57.

2. *CofE*; Sargent and Bottin, *Case Histories*, 17; McComb, *Galveston*, 61.

3. Wolff, *Indianola and Matagorda Island*, 69; Sargent and Bottin, *Case Histories*, 22; James P. Baughman, "The Evolution of Rail-Water Systems of Transportation in the Gulf Southwest, 1836–1890," *Journal of Southern History* 34 (1968): 370–71; Baughman, *Charles Morgan*, 199–200; Mary McAdams Sibley, *The Port of Houston: A History* (Austin: University of Texas Press, 1968), 19, 89–99.

4. "Dugout" was the common name for "natural channels enhanced by man." See Coffee, "After 90 Years, 5 Old Diaries Make News Again," *Corpus Christi Times*, August 5, 1957; and Ford, *Mercer Logs*, 54. The indentation was also known as the Morris and Cummings Cut. See John Dunn, interview, August 31, 1929, Corpus Christi, TX, Paul Schuster Taylor Papers, Bancroft Library, University of California, Berkeley (Banc MCS 84/38c).

5. "Letter from Austin," "The Aransas Road Company," January 28, 1860; "The Amended Charter of Corpus Christi," "Austin Correspondence," February 25, 1860, all from the *Ranchero* (Corpus Christi, TX); Delaney, *Maltby Brothers' Civil War*, 73.

6. *Centennial History of Corpus Christi*, 88; Phyllis Coffee, "Family Has Stayed for 6 Generations," *Corpus Christi Times*, August 7, 1957.

7. Baughman, *Charles Morgan*, 188–90, 201–8; Baughman, "Evolution of Rail-Water Systems," 371; Sibley, *Port of Houston*, 99–100.

8. George Cooper, "The Railroads of South Texas," and John Lloyd Bluntzer, "The Texas-Mexican Railroad" (presentations at the South Texas Historical Association, Rockport, TX, April 6, 2019); Cheeseman, "Richard King"; George C. Werner, "Texas Mexican Railway," *Handbook of Texas Online*, accessed December 31, 2017, http://www.tshaonline.org/handbook/online/articles/eqt21; J. L. Allhands, "Lott, Uriah," *Handbook of Texas Online*, accessed December 31, 2017, https://www.tshaonline.org/handbook/entries/lott-uriah.

9. *Map of San Antonio and Aransas Pass Railway and Connections* (San Antonio, TX: Mills Engineering, n.d.), in author's possession; "S.A. and A.P.R.R," March 21, 1886; "The S.A. and A.P. Railroad," May 2, 1886; "Commissioner's Court," August 15, 1886; "Gossip and Taffy," November 7, 1886; "The First Excursion," November 14, 1886, all from *Corpus Christi Caller*.

10. The values in 2017 currency are $2,492.02, $4,734.84, and $124.68, respectively. See inflation calculator, Official Inflation Data, Alioth Finance, accessed December 30, 2017, http://www.in2013dollars.com/1886-dollars-in-2017?amount=5; "The S.A. and A. P. Railroad," May 2, 1886; "From Corpus Christi: A Texas Town That Will Soon Have a Boom," May 2, 1886; "Corpus Christi: The Seaside Attractions of a Lovely Bay City," August 27, 1887, all from *Corpus Christi Caller*.

11. McComb, *Galveston*, 59–61; Sibley, *Port of Houston*, 114; Thompson, "Governmental Responses," 46.

12. "Is the Galveston News Mad?," *Corpus Christi Caller*, October 13, 1886; Sibley, *Port of Houston*, 114–25; Sargent and Bottin, *Case Histories*, 22.

CHAPTER 11

1. Werner, "Texas Mexican Railway."

2. "S.A.A.P." was the local term for the San Antonio and Aransas Pass Railway.

3. "A Field for Enterprise," November 14, 1886; "A Trip to Padre Island," April 9, 1887; untitled, June 25, 1887; "The Coast Outlets," August 27, 1887, all from *Corpus Christi Caller*.

4. Williams, "Star of St. Mary's."

5. "Corpus Christi: The Seaside Attractions of a Lovely Bay City," "The Coast Outlets," August 27, 1887, both from *Corpus Christi Caller*.

6. Francaviglia, *From Sail to Steam*, 241; Keith Guthrie, "Aransas Pass, TX," *Handbook of Texas Online*, accessed July 24, 2017, http://www.tshaonline.org/handbook/online/articles/hfa06; Keith Guthrie, "Aransas Harbor Terminal Railway," *Handbook of Texas Online*, accessed August 30, 2017, https://www.tshaonline.org/handbook/entries/aransas-harbor-terminal-railway.

7. Coffee, "Logs Reveal Texas Gulf Coast History," 229; Ford, *Mercer Logs*, 55; Stewart, "History of the Aransas Pass Jetties"; *U.S. Coast Survey: Corpus Christi*

Pass, 1869, chart courtesy of PAPHA; "Map of Corpus Christi Bay and Aransas Pass," *Corpus Christi Caller*, January 1, 1887, courtesy of the University of Texas Marine Science Institute.

8. William Kent, "Ropes Pass, Texas," *Engineering Magazine* 3 (June 1892): 342, courtesy of PAPHA; Committee on Publications, "Discussions on Harbors," *Transactions of the American Society of Civil Engineers*, vol. 54, part A (New York: American Society of Civil Engineers, 1905), 403, hereafter cited as *Transactions*.

9. Committee on Publications, "Discussions on Harbors."

10. O. H. Ernst, Major of Engineers, to the Chief of Engineers, USA, June 22, 1887, *CofE*, 1312–13.

11. Stewart, "History of the Aransas Pass Jetties"; *Centennial History of Corpus Christi*, 88; John Dunn, interview.

12. Tunnell et al., *Encyclopedia of Texas Seashells*, 89; Stewart, "History of the Aransas Pass Jetties"; Guthrie, "Aransas Pass, TX"; *Improvement of Aransas Pass*, 1431; Committee on Publications, "Discussions on Harbors," 442; *Aransas Pass Texas Made under the Direction of Bvt. Lieut. S. M. Mansfield, May 21st to 28th 1885*, chart courtesy of PAPHA; *Aransas Pass—1895*, chart courtesy of PAPHA.

13. "The Texas Appropriations," April 18, 1886; "From Corpus Christi: A Texas Town That Will Soon Have a Boom," May 2, 1886, both from *Corpus Christi Caller*; *Improvement of Aransas Pass*, 1432.

14. Ernst to Chief of Engineers, *CofE*, 1317.

15. *Improvement of Aransas Pass*, 1431–32; Stewart, "History of the Aransas Pass Jetties"; "Engineer's Report on Aransas Pass," *Corpus Christi Caller*, September 3, 1887.

16. Report of Mr. N. E. Savage, Assistant Engineer, to Major O. H. Ernst, June 30, 1888, *CofE*, 1308–12; Sargent and Bottin, *Case Histories*, 37.

17. Francaviglia, *From Sail to Steam*, 241, 243–44; "Aransas Pass, Texas," in *The Handbook of Texas*, ed. Walter Prescott Webb (Austin: Texas State Historical Association, 1952); Savage to Ernst, *CofE*, 1310; Stewart, "History of the Aransas Pass Jetties"; Thompson, "Governmental Responses," 46; Sibley, *Port of Houston*, 114; "Plans for the Harbor," *Galveston Daily News*, April 25, 1900; Thomas Lincoln Casey to the Chief of Engineers, July 19, 1887, *CofE*, 1318–19.

18. The approximate site of dredging was eleven miles south of Aransas Pass and a little north of Mustang Island's Water Exchange Pass. See *Nueces County Highway Map* (Nueces County Department of Public Works, 1996), and Kent, "Ropes Pass," 3–5.

19. Alan Lessoff, *Where Texas Meets the Sea: Corpus Christi and Its History* (Austin: University of Texas Press, 2015), 99–100; Givens and Moloney, *Corpus Christi*, 144–46.

20. J. Guthrie Ford, *A Texas Island* (Port Aransas, TX: USA Hurrah Publishing, 2008), 24.

21. "Port Aransas, Past, Present, Future," *Port Aransas Post*, November 24, 1911.

22. Miller Harwood and W. A. Serivner, *Fabulous Port Aransas* (1949), part 3, Travel South Texas, accessed July 25, 2017, www.stxmaps.com/go/fabulous-port-aransas3.html; Stewart, "History of the Aransas Pass Jetties"; "Aransas Pass, Texas," 57.

23. Ernst to Chief of Engineers, *CofE*, 1315–17; Stewart, "History of the Aransas Pass Jetties"; Sargent and Bottin, *Case Histories*, 37.

24. Harwood and Serivner, *Fabulous Port Aransas*; Stewart, "History of the Aransas Pass Jetties"; "Report of the Chief of Engineers, US Army," *Annual Reports of the War Department for the Fiscal Year Ended June 30, 1901*, part 3 (Washington, DC: Government Printing Office, 1901), 1956; Committee on Publications, "Discussions on Harbors," 341, 389.

25. Committee on Publications, "Discussions on Harbors," 392–94.

26. Tunnell and Judd, *Laguna Madre*, 23; Pilkey, *Celebration of the World's Barrier Islands*, 46; Committee on Publications, "Discussions on Harbors," 386, 403, 409, 428, 431, 432–33.

27. Stewart, "History of the Aransas Pass Jetties"; Committee on Publications, "Discussions on Harbors," 415–16, 419–21, 444–45.

28. Ford, *Texas Island*, 16; Committee on Publications, "Discussions on Harbors," 416; *Aransas Pass, Texas, from Survey made by Lieut. Col. S. M. Mansfield, 1885*, chart courtesy of PAPHA; Committee on Publications, "Discussions on Harbors," 442–43; Sargent and Bottin, *Case Histories*, 37; Stewart, "History of the Aransas Pass Jetties."

29. Tunnell et al., *Encyclopedia of Texas Seashells*, 36–48; Stewart, "History of the Aransas Pass Jetties"; Montagna et al., *Characterization of Anthropogenic and Natural Disturbance*, 30, 32; Barney Farley, *Fishing Yesterday's Gulf Coast* (College Station: Texas A&M University Press, 2002), 25; Horace Logan Whitten, "Marine Biology of the Government Jetties in the Gulf of Mexico Bordering the Texas Coast" (master's thesis, University of Texas, 1940), University of Texas at Austin: Texas Scholar Works, accessed January 8, 2018, https://repositories.lib.utexas.edu/handle/2152/22109.

30. Untitled, *Galveston Daily News*, January 12, 1896; Sargent and Bottin, *Case Histories*, 37–38; Tunnell et al., *Encyclopedia of Texas Seashells*, 54; Stewart, "History of the Aransas Pass Jetties."

31. *Aransas Pass, 1895, Survey of H. C. Ripley*, chart, *Transactions*, fig. 38; *Aransas Pass, August 28, 1896, Survey by the Aransas Pass Harbor Company*, chart, *Transactions*, fig. 39; *Aransas Pass, February 2, 1897, Survey by the Aransas Pass Harbor Company*, chart, *Transactions*, fig. 40, all on 423.

32. Stewart, "History of the Aransas Pass Jetties."

CHAPTER 12

1. "Aransas Pass' Motto: Where Everybody Shall Make Money," *Wise County Messenger* (Decatur, TX), January 31, 1896; "Deep Water for Texas," *Bryan Daily Eagle*, January 31, 1896; untitled, *Shiner Gazette*, February 18, 1896.
2. "Aransas Harbor Bill," *Brownsville Daily Herald*, January 18, 1896; "Foolish Fakirs Filed," *Victoria Advocate*, February 29, 1896; "Remarkable Results," *San Antonio Light*, April 2, 1896.
3. Untitled, "Dismal Failure," September 10, 1896, both from *Brownsville Daily Herald*.
4. Untitled, *Brownsville Daily Herald*, September 18, 1896; "Report of the Chief of Engineers," part 2, 1668, 1672; "The Dynamite Effective," *Brownsville Daily Herald*, December 9, 1896; "Work on Deep Water," *San Antonio Light*, September 30, 1896; "To Resume Work," *Galveston Daily News*, November 20, 1896.
5. "Report of the Chief of Engineers," part 2, 1668–69; "The Dynamite Effective"; untitled, *Shiner Gazette*, January 1, 1897; "For Deep Water," *San Antonio Light*, February 16, 1897; untitled, December 16, 1896; December 26, 1896; March 2, 1897, all from *San Antonio Light*; "Aransas Pass Company," March 1, 1897; "Government Engineers," "Representative from Corpus Christi," July 22, 1897; "The Coast Region of Texas," December 22, 1897, all from *Galveston Daily News*.
6. "Report of the Chief of Engineers," part 2, 1669–73; "Aransas Harbor Work," *Brownsville Daily Herald*, June 6, 1897; "Government Engineers," *Galveston Daily News*, July 22, 1897.
7. "Telegrams Briefed," *Daily San Antonio Light*, December 10, 1897; "Aransas Pass Work," *Shiner Gazette*, December 15, 1897; "*The Sun's* Silly Attack," *Brownsville Daily Herald*, May 4, 1898; "Neighborhood Notes," *El Paso Daily Herald*, December 16, 1898; "At Aransas Pass," *Corpus Christi Caller*, November 24, 1899; "For Aransas Pass," *Victoria Daily Advocate*, June 20, 1901; "Work on Aransas Bay," *Corpus Christi Caller*, November 6, 1901; "After a Long Hearing," *Galveston Daily News*, March 24, 1903.
8. "One Jetty Plan," *Galveston Daily News*, April 22, 1904.
9. "One Jetty Plan."
10. Whitten, "Marine Biology"; Tunnell et al., *Encyclopedia of Texas Seashells*, 48–52, 57–59, 210, 298; Henry Tan, "Underwater Explosion," University of Aberdeen, 2008, accessed August 19, 2018, https://homepages.abdn.ac.uk/h.tan/pages/teaching/explosion-engineering/Underwater-I.pdf; Jessica Macdonald, "The Damage of Dynamite Fishing," Scuba Diver Life, September 1, 2014, accessed August 21, 2018, https://scubadiverlife.com/damage-dynamite-fishing.

11. R. B. Talfor to Captain C. S. Riché, December 14, 1900, "Report to the Chief of Engineers," part 3, 1954–55; "One Jetty Plan."

12. "One Jetty Plan"; "Captain Edward T. Mercer Jr.," Find a Grave, accessed August 26, 2018, https://www.findagrave.com/memorial/53334663/edward-thomas-mercer#clipboard; Crnkovich, "Robert Ainsworth Mercer," 9; Mercer Papers, courtesy of PAPHA; Committee on Publications, "Discussions on Harbors," 434; Stewart, "History of the Aransas Pass Jetties"; "One Jetty Plan"; Report of F. Oppikoper, "Appendix U—Report of Captain Jadwin"; context indicates this is from "Report of the Chief of Engineers, US Army," *Annual Reports of the War Department* (Washington, DC: Government Printing Office, 1904), 2009, courtesy of PAPHA. Haupt later sued the US government for failing to remunerate him adequately, but the Supreme Court upheld the Court of Claims ruling that his claim was unjustified. See *Lewis V. Haupt v. United States in Cases Argued and Decided in the Supreme Court of the United States, October Term, 1920* (Rochester, NY: Lawyers' Cooperative, 1922), 272.

13. *Entrance to Aransas Pass Showing Shore Lines from 1851 to 1904*, chart, *Transactions*, fig. 48; "Jetty Repair Contracts," July 30, 1905; "Like Capt. Jadwin," September 9, 1905; "Harbor Company, in 1899, Transferred All Its Holdings to the Government," November 22, 1906, all from *Galveston Daily News*.

14. "Will Attend," July 7, 1905; "Convention Matters," July 25, 1905, both from *Daily Victoria Advocate*; "Like Capt. Jadwin."

15. "Work at Aransas Pass," *Shiner Gazette*, January 12, 1898.

16. Lynn Alperin, *History of the Gulf Intracoastal Waterway*, Navigation History NWS-83-9 (January 18, 1983), 4; "Waterways Improvement a Paramount Necessity," *Brownsville Daily Herald*, April 25, 1906.

17. "Aransas Pass Lands Titles Perfected," September 13, 1907, *San Antonio Light*; "Government Work at Aransas Pass," *Victoria Advocate*, September 12, 1908; untitled, *Brackett News Mail*, July 15, 1910; Sargent and Bottin, *Case Histories*, 37–39; Stewart, "History of the Aransas Pass Jetties."

18. "Government Work at Aransas Pass"; Sargent and Bottin, *Case Histories*, 39.

19. Stewart, "History of the Aransas Pass Jetties"; untitled, *Galveston Daily News*, January 12, 1896; Sargent and Bottin, *Case Histories*, 38–39; "Aransas Pass: Migration and Jetties" (poster display, courtesy of PAPHA).

20. "Government Work at Aransas Pass," *Victoria Advocate*, September 12, 1908; Ernst to Chief of Engineers, *CofE*, 1315; Harwood and Serivner, *Fabulous Port Aransas*; "Want Deep Water," *Victoria Weekly Advocate*, April 17, 1909; Stewart, "History of the Aransas Pass Jetties."

CHAPTER 13

1. "Red Drum (*Sciaenops ocellatus*)," Texas Parks and Wildlife, accessed September 20, 2018, https://tpwd.texas.gov/huntwild/wild/species/reddrum/.

2. "The Largest Unassailable, Commercial, and Strategic Location of Rockport," February 27, 1910, *San Antonio Light and Gazette*, and "The Largest Land-Locked Harbor on the Gulf Coast," copy, courtesy of PAPHA; "History," *Port Aransas South Jetty*, February 2, 1983; "Rockport, TX, April 19, 1908" in *Map of Public Land on Mustang Island*, courtesy of PAPHA.

3. "Deep Water Channel to Be Surveyed," *Corpus Christi Caller*, September 3, 1909; "Rivers and Harbors Bill," *Bonham Daily Favorite*, February 11, 1910; "Waterfront Bill Now before the Legislature," *Corpus Christi Caller*, May 21, 1915; "Survey of 25-Foot Channel is Assured," *Corpus Christi Caller*, January 27, 1909.

4. "Steam Dredges," *Steam Shovel and Dredge* 13 (November 1909): 703; "Aransas Pass, Texas"; Guthrie, "Aransas Harbor Terminal Railway"; "History"; "For Deep Water at City of Aransas Pass," *Galveston Daily News*, May 6, 1909.

5. "For Deep Water at City of Aransas Pass."

6. "For Deep Water at City of Aransas Pass."

7. "Channel to Rockport," January 7, 1910; "Board of Engineers Assembles Here," November 5, 1910, both from *Galveston Daily News*; "Want Deep Water," *Victoria Weekly Advocate*, April 17, 1909.

8. "Joseph Hirsch to the Bankers at Corsicana," *Weekly Corpus Christi Caller*, February 28, 1908; "Survey of 25 Foot Channel Is Assured," *Corpus Christi Caller*, January 27, 1909; untitled, *Corpus Christi Caller*, September 14, 1900; "Corpus Christi's Advantages," *Corpus Christi Caller*, Christmas Souvenir Ed., December 2, 1901.

9. "Port of Aransas Pass," *Galveston Daily News*, March 6, 1910; "Dock and Wharf Company to Build Connecting Railroad," *Laredo Weekly Times*, March 6, 1910; "Aransas Pass Project Will Be Financed," *San Antonio Light*, September 14, 1911.

10. "Aransas Pass on the Mainland," *San Antonio Light and Gazette*, May 22, 1920; "Aransas Pass: City Lot Contracts," *Wise County Messenger* (Decatur, TX), December 10, 1908.

11. "The Need of a Seaport at Aransas Pass," *San Antonio Light and Gazette*, March 13, 1910.

12. "Honorable J. L. Slayden at Corpus Christi," *Galveston Daily News*, October 22, 1909.

13. "Aransas Pass," *Rockdale Reporter and Messenger*, October 13, 1910; "Aransas Pass Makes Many Improvements," *San Antonio Light*, October 18, 1910.

14. "Port of Aransas Pass Is Opened," *San Antonio Light and Gazette*, October

20, 1910; "Celebrate Completion of the Ship Channel," *Galveston Daily News*, October 21, 1910.

15. "Port Aransas, Past, Present, Future," *Port Aransas Post*, November 24, 1911.

16. "Board of Engineers Assembles Here November 21," *Galveston Daily News*, November 5, 1910; "Aransas Pass, Texas." In 2016 currency, these sums would be equivalent to $85,897,242.11 for the channel to Corpus Christi, $50,527,789.47 for the channel to Rockport, and $15,158,336.84 for the channel to Aransas Pass. See inflation calculator, Official Inflation Data, Alioth Finance, accessed September 26, 2019, http://www.in2013dollars.com/1910-dollars-in-2016?amount=3400000.

17. "Engineers Report on Aransas Pass Project," *Galveston Daily News*, December 14, 1910. In 2016 currency, that sum would be equivalent to $9,473,960.50. See inflation calculator, Official Inflation Data, Alioth Finance, accessed September 26, 2018, http://www.in2013dollars.com/1910-dollars-in-2016?amount=3400000.

18. "Our Future Greatness Assured," December 15, 1911; "Meet Me at Port Aransas," December 8, 1911; "$250,000 for the Lion's Portion," November 24, 1911; "Incorporated under Commission Form," December 1, 1911; all from *Port Aransas Post*.

19. "Another Port for West Texas," *Galveston Tribune*, May 5, 1911; "Aransas Pass Project Will Be Financed."

20. "Rockport to the Defense of Its Harbor," *San Antonio Light*, April 11, 1912.

21. "Port of Aransas Pass," *Galveston Daily News*, March 6, 1910; "Operations at Aransas Pass Make Headway," *San Antonio Light*, September 24, 1911; "*Comstock* Has Serious Accident," *Port Aransas Post*, January 26, 1912; *Map of the City of Aransas Pass, Texas*, G. P. Tarrant, comp., n.d., copy in author's possession; "Letter from the Secretary of War," US Congress, House Documents, 62nd Cong., 3rd sess., Congressional Serial Set, issue 6393, vol. 27, part 2, December 2, 1912–March 4, 1913 (Washington, DC: Government Printing Office, 1913), 46, 50, 52. Estimated dredging costs ran to $250,000 at the time, the equivalent of $6,185,000.00 in 2016. See inflation calculator, Official Inflation Data, Alioth Finance, accessed November 20, 2018, http://www.in2013dollars.com/1912-dollars-in-2016?amount=250000.

22. "Screwmen, Spidermen, and Cotton's Gilded-Age Gargantua," The History Bandits, accessed November 20, 2018, https://thehistorybandits.com/2015/02/13/screwmen-spidermen-and-cottons-gilded-age-gargantua/; "Port Aransas, Texas," in *Water Terminal and Transfer Facilities: Letter from the Secretary of War* (Washington, DC: Government Printing Office, 1921), 1142–43.

23. "Aransas Pass Project Will Be Financed," "Aransas Pass Project to Be Finished Soon," both from *San Antonio Light*, September 17, 1911.

24. "Army Engineers' Report on Port Aransas," *San Antonio Light*, December 22,

1912; "Shipping" and "Barge Line Facilities," both from *Aransas Pass Texas: Where Sails Meet Rails* (Aransas Pass: Chamber of Commerce, August 1913), accessed October 20, 2018, www.larryray.com/aransas-pass-tx.html; Frankl and Garcia, *Soil Survey*, 11–13.

25. "Shipping"; Lawrence Dunn, *The World's Tankers* (London: Adlard Coles, 1956), 33–36, 52–56.

26. By June 1919, Aransas Pass was 104 feet wide at its narrowest point and over 22 feet deep. See *War Department Annual Reports, 1919*, vol. 2, *Report of the Chief of Engineers* (Washington, DC: Government Printing Office, 1919), 1115.

27. Francaviglia, *From Sail to Steam*, 257–59; "Tugboats and Towboats," Tugboat Enthusiasts Society of the Americas, accessed December 16, 2018, http://www.tugboatenthusiastsociety.org/Pages/tugboats-and-towboats-01.htm (site discontinued; printout in author's possession); Selective Service Registration Card #1309, US World War I Draft Registrations, Ancestry, accessed December 13, 2018, https://search.ancestry.com/search; fuel receipt, January 14, 1913, and check, April 15, 1916, both courtesy of PAPHA; "*Comstock* Has Serious Accident"; "Oil Barge In," *Port Aransas Post*, December 15, 1911; *Fifteenth Census of the United States, 1930*, Records from the Bureau of the Census.

28. Edward Burr, Colonel, Corps of Engineers, to Secretary of War, December 4, 1912, US Congress, House Documents, 62nd Cong., 3rd sess., December 2, 1912–March 4, 1913, 51; "A Port Aransas 'Pioneer,'" *San Antonio Light*, October 22, 1911; J. L. Terrell and James A. Cook, "Magnolia Petroleum Company," *Handbook of Texas Online*, accessed December 13, 2018, http://www.tshaonline.org/handbook/online/articles/dom01.

29. "Army Engineers' Report on Port Aransas"; "Shipping"; "Port Aransas, Texas," 1143.

CHAPTER 14

1. Holt, *State of the Bay*, 42; Tunnell and Judd, *Laguna Madre*, 175, 185–86.

2. "Texas Newspaper Comment," *Galveston Daily News*, December 13, 1896; "Tarpon Island and Aransas Pass, Tarpon, Texas," *Galveston Daily News*, September 26, 1898; Harwood and Serivner, *Fabulous Port Aransas*, part 4.

3. Farley, *Fishing Yesterday's Gulf Coast*, 21–22; Tunnell and Judd, *Laguna Madre*, 226–29.

4. Also known as jewfish, its name was changed to goliath grouper by the Atlantic Fisheries Society in 2001. See Avishay Artsy, "How the Jewfish Got Its Name," Jewish Telegraphic Agency, https://www.jta.org; "Goliath Grouper: Fish,"

Encyclopedia Britannica, https://www.britannica.com, both accessed December 22, 2019; "Greatness of the Tarpon," August 15, 1897; "Tarpon Island and Aransas Pass, Tarpon, Texas," September 26, 1898, both from *Galveston Daily News*.

5. Because of their swim bladder, which "acts as a lung so they can breathe raw air," tarpon are able to take in an unusual amount of oxygen. See "Tarpon Facts," TarponFish.com, accessed December 23, 2018, https://www.tarponfish.com; "Greatness of the Tarpon"; Farley, *Fishing Yesterday's Gulf Coast*, 40–43.

6. "Greatness of the Tarpon"; Farley, *Fishing Yesterday's Gulf Coast*, 43–46; "San Antonians at Corpus Christi," *San Antonio Light*, July 31, 1898.

7. J. Guthrie Ford and Mark Creighton, *Images of America: Port Aransas* (Charleston, SC: Arcadia Publishing, 2010), 92–93; Ford, *Texas Island*, 22–24; Phil H. Shook, "Farley Boast and Tarpon: The Farley Family Boatbuilders," *Texas Parks and Wildlife*, October 1995; "Local Guide Farley Dies," *Port Aransas South Jetty*, n.d., both courtesy of PAPHA; "Additional City News," *Port Aransas Post*, January 26, 1912; ads placed in *Port Aransas Post*, December 15, 1911; "Incorporated under Commission Form, " December 1, 1911, *Port Aransas Post*; "Letter from the Secretary of War," 62nd Cong., 3rd sess., 54; "Port Aransas Again Favored," *Port Aransas Post*, January 26, 1912; Joe Holley, "The Truth about Saloons at Sea, Dune-Loving Cayotes," *Houston Chronicle*, October 8, 2016.

8. "Industries" and "Directory," both from *Aransas Pass Texas: Where Sails Meet Rails*; John Guthrie Ford, "Harbor Island Once Bustling," *Port Aransas South Jetty*, July 11, 2013; "Poor Little Port Aransas," *San Antonio Light*, November 8, 1913.

9. "Hunting and Fishing," in *Aransas Pass Texas: Where Sails Meet Rails*; "Engineering News," *Engineering News Record*, June 1, 1911; Kratz, "History of the Texas Shrimp Industry," 29.

10. "Aransas Pass," in *Aransas Pass Texas: Where Sails Meet Rails*.

11. Atlee M. Cunningham, *Corpus Christi Water Supply Documented History: 1852–1997* (Corpus Christi, 1997), 10, 17, 47, 70, Local History Room, Corpus Christi Public Library, hereafter cited as LHR; "Citizens' Committee Is Appointed to Assist in Waterfront Improvement," *Corpus Christi Caller*, December 28, 1918; "Burk Oil Company Opens an Office Here," March 5, 1919; "Nueces County Oil Fields to Be Developed," March 13, 1919; "Lucky Jim Oil Company Opens Office in City," April 9, 1919, all from *Corpus Christi Caller*.

12. "San Antonians at Corpus Christi"; "Corpus Christi Gets Plant of Big Concern," *Corpus Christi Caller*, May 14, 1919; Margaret Patrice Slattery, *Promises to Keep: A History of the Sisters of Charity of the Incarnate Word*, vol. 2 (San Antonio: Sisters of Charity of the Incarnate Word, 1995), 386n3; "Rest Camp to Be Enlarged Is Indication," August 22, 1919; "Proposal to Celebrate the Opening of the Causeway,"

June 28, 1915; "U. S. Will Help Dredge Bay at Corpus Christi," September 11, 1919, all from *Corpus Christi Caller*.

13. *Thirteenth Census of the United States, 1910, Texas*, vol. III, *Nueces County*, Records from the Bureau of the Census, LHR; "Deep Water Is Vital for Corpus Christi Prosperity," *Corpus Christi Caller*, June 6, 1919; Anita Eisenhauer and Gigi Starnes, *Corpus Christi, Texas: A Picture Postcard History* (Corpus Christi, TX: Anita's Antiques, 1987), 87–89; *Corpus Christi, Texas: June 1914* (New York: Sanborn Map Company, 1914); *US Coast and Geodetic Chart*, 1923, courtesy of PAPHA; "Deep Water to Be Had If People Pull Together," *Corpus Christi Caller*, August 25, 1916; Ford, *Texas Island*, 23; "Down to Work," *Corpus Christi Caller*, February 21, 1919.

14. "Waterfront Bill Now before the Legislature," May 21, 1915; "Deep Water Is Vital for Corpus Christi Prosperity," "Rivers and Harbors Committee to Meet in Capitol Next Month," January 21, 1919; "Hard Surface Road to Corpus Christi Is an Imperative Need," August 16, 1918; "Roy Miller to Manage Union War Fund Drive to Start November 11," October 16, 1918; "City News in Brief," November 19, 1918; all from *Corpus Christi Caller*; Givens and Moloney, *Corpus Christi*, 207.

15. "War Department Orders Engineers Here to Aid City in Seawall Plans," January 14, 1919; "City's Waterfront Plan Is Favored," February 15, 1919; "Waterfront Fight Ends," February 16, 1919; "Mr. Voter," March 29, 1919, all from *Corpus Christi Caller*; "Some Waterfront History," *Corpus Christi Times*, August 12, 1918.

16. "Terse Tales of Interest from Thriving Towns," February 19, 1911; "Rockport Man Enthusiastic," March 30, 1911, both from *San Antonio Light and Gazette*; "Entire Ticket Headed by Judge Gordon Boone Swept into Office in City," *Corpus Christi Caller*, April 2, 1919. For the full story of that 1919 mayoral election, see Mary Jo O'Rear, *Storm over the Bay: The People of Corpus Christi and Their Port* (College Station: Texas A&M University Press, 2009).

17. "Rockport to the Defense of Its Harbor," *San Antonio Light*, April 11, 1912; "Rockport!," *San Antonio Light and Gazette*, February 19, 1911.

18. "Our Deep Water Port," *San Antonio Light*, April 8, 1914; Farley, *Fishing Yesterday's Gulf Coast*, 14; Krantz, "History of the Texas Shrimp Industry," 29, 37; "Captain V. A. (Brother) Court, Stricken while Boarding a Tanker in the Gulf," unknown news source, November 3, 1954, courtesy of PAPHA; "And They Went to Rockport," *Shiner Gazette*, September 10, 1896; Kam Wagert and Pam Stranahan, *Aransas County in Postcards* (Fulton, TX: Friends of the History Center for Aransas County, 2014), 57; William Allen and Sue Hastings Taylor, "Aransas: The Life of a Texas Coastal County," vol. 2 (unpublished manuscript, Aransas County Historical Society, 1997), 339–41; Pam Stranahan, "Why Was WWI Called 'The Great War'?," History Center for Aransas County, accessed October 5, 2018, https://www.thehistory

centerforaransascounty.org/history-mystery-1/why-was-wwi-called-%E2%80%98 the-great-war%E2%80%99%3F.

19. *War Department Annual Reports, 1919*, 1116–17; *War Department Annual Reports, 1923, Report of the Chief of Engineers*, part 1 (Washington, DC: Government Printing Office, 1923), 1115–17.

20. "Ten Concrete Ships Now Building at Port Aransas," *Concrete and Engineering News* 31 (January 1919): 48, courtesy of PAPHA; Dunn, *World's Tankers*, 66; "Concrete Tanker Built Like Grain Elevator," *Popular Mechanics* 24 (November 1920): 695.

21. "Merchant Marines during World War II" (display at the National World War II Museum, New Orleans, LA, November 2017); "Navigation in the 19th to 20th Centuries," Penobscot Marine Museum, https://penobscotmarinemuseum.org/pbho-1/history-of-navigation/navigation-19th-20th-centuries; "Stadimeter," Smithsonian National Museum of American History, accessed July 19, 2019, https://amhistory.si.edu/navigation/type.cfm?typeid=13; "History of Navigation at Sea," Water Encyclopedia, accessed July 18, 2019, http://www.waterencyclopedia.com/Mi-Oc/Navigation-at-Sea-History-of.html.

22. Thomas J. White, *United States Early Radio History* (blog), https://earlyradiohistory.us/sec005.htm; Jerry Proc, "A Brief History of Naval Radio Communications," http://jproc.ca/rrp/nro_his.html; Amitava Chakrabarty, "What Marine Communication Systems Are Used in the Maritime Industry?," Marine Insight, https://www.marineinsight.com/marine-navigation/marine-communication-systems-used-in-the-maritime-industry/, all accessed July 18, 2019.

23. Dunn, *World's Tankers*, 66; Allen and Taylor, "Aransas," 332–33; Norman Frank, "Shipyards in Rockport," Historical Marker Database, accessed October 5, 2018, https://hmdb.org/marker.asp?marker=58824.

24. Stranahan, "Why Was WWI Called 'The Great War'?"; Allen and Taylor, "Aransas," 335–38.

CHAPTER 15

1. *Soil Survey of Padre Island*, 10–11; Nummedal, *Sedimentary Processes*, 8; Varnum, Álvar Núñez *Cabeza de Vaca*, 51; Shorey, *Story of the Maine Fifteenth*, 54–55, 62–65; Committee on Publications, "Discussions on Harbors," 430; Bruseth and Turner, *From a Watery Grave*, 5; "Celebration Completion of Ship Channel," *Galveston Daily News*, October 21, 1910.

2. "Anatomy of a Hurricane," *Corpus Christi Caller-Times*, September 15, 1988; "What Causes Hurricanes?," WeatherQuestions.com, accessed February 10, 2019, www.weatherquestions.com; Ford, *Mercer Logs*, 65; Wolff, *Indianola and*

Matagorda Island, 66–77; David Roth, "Texas Hurricane History" (unpublished manuscript, National Weather Service, 2000), 14–15, 17. It was to forestall such rampant gouging that, along with Harbor Basin, the army ordered a ten-thousand-foot-long dike to be built atop St. Joseph's in 1910.

3. Brownson Malsch, *Indianola: The Mother of West Texas* (Austin: State House Press, 1988), 91, 262; Wolff, *Indianola and Matagorda Island*, 69, 74–76; "Indianola and Galveston," *Corpus Christi Caller*, August 29, 1886.

4. "Is the *Galveston News* Mad?," *Corpus Christi Caller*, October 13, 1886; Roth, "Texas Hurricane History," 21.

5. For a discussion of hurricanes that hit Galveston Bay between 1818 and 1886, see Roth, "Texas Hurricane History," 12, 17, 22–23; and McComb, *Galveston*, 27–31, 220.

6. "Galveston Gossip," *Corpus Christi Caller*, June 20, 1886; Patricia Bellis Bixel and Elizabeth Hayes Turner, *Galveston and the 1900 Storm: Catastrophe and Catalyst* (Austin: University of Texas Press, 2000), 17; Rod Beemer, *The Deadliest Woman in the West: Mother Nature on the Prairies and Plains, 1800–1900* (Caldwell, ID: Caxton Press, 2006), 317; I. Cline, "West India Hurricanes," *Galveston Daily News*, July 6, 1891.

7. "Great Disaster at Galveston," September 10, 1900; "Number of Dead May Reach 10,000," September 11, 1900, both from *New York Times*; "Galveston's Calamity," September 14, 1900; "A Visit to Galveston," September 21, 1900, both from *Corpus Christi Caller*; McComb, *Galveston*, 122. Within the city at least six thousand died and over three thousand homes were destroyed; property damages throughout Galveston County totaled $5.5 million. See Isaac M. Cline, "Special Report on the Galveston Hurricane of September 8, 1900," National Weather Service Heritage, accessed December 1, 2003, https://vlab.noaa.gov/web/nws-heritage/-/galveston-storm-of-1900. This event has been cited as "the worst disaster in the history of the United States." See Gary Cartwright, *Galveston: A History of the Island* (Fort Worth: Texas Christian University Press, 1991), 180; and John Edward Weems, *A Weekend in September* (College Station: Texas A&M University Press, 1957), 167.

8. "Galveston's Calamity," "An Appeal," "Horror of Horrors," September 14, 1900; "Dead Animals on Mustang Island," "Captain Mercer's Experiences," September 21, 1900, all from *Corpus Christi Caller*.

9. "Where Is Port Aransas?," January 26, 1912; "Get Here before the Railroads Do!," December 15, 1911, both from *Port Aransas Post*; "Poor Little Port Aransas," *San Antonio Light*, November 8, 1913.

10. "The Largest Landlocked Harbor on the Gulf Coast," *San Antonio Light and Gazette*, February 27, 1910; "Want Deep Water," *Victoria Weekly Advocate*,

April 17, 1909; "Celebrate Completion of the Ship Channel," *Galveston Daily News*, October 21, 1910; "No Place in the Entire South Has More Brilliant Prospects," *San Antonio Light*, August 6, 1911.

11. "Corpus Christi as a Safe Harbor and Deep Water Port," July 30, 1909; George Reeder, "Corpus Christi's Advantages," December 20, 1901, both from *Corpus Christi Caller*.

12. "City and Country," June 2, 1888; "Indianola and Galveston," September 21, 1900; Reeder, "Corpus Christi's Advantages," all from *Corpus Christi Caller*; Kent, "Ropes Pass, Texas."

13. W. L. Coleman, "A Great Health Resort," December 28, 1900; untitled, September 14, 1900; Reeder, "Corpus Christi's Advantages," all from *Corpus Christi Caller*.

14. Reeder, "Corpus Christi's Advantages."

15. Cline, "Special Report on the Galveston Hurricane"; McComb, *Galveston*, 138–43; Cartwright, *Galveston*, 190; "Pilot Boy in the Storm," *Corpus Christi Caller*, July 30, 1909. The cost was $1,250,000, equivalent in 2016 to $34,091,903.41. See inflation calculator, Official Inflation Data, Alioth Finance, accessed February 10, 2019, http://www.in2013dollars.com/1905-dollars-in-2016?amount=1250000.

16. "Work on Texas Waterways," *Galveston Daily News*, July 21, 1910; Roth, "Texas Hurricane History," 31; "Storm in the Gulf," October 16, 1912; "A Severe Storm Visited Brownsville Yesterday," "Hurricane Flag Raised in Corpus," October 17, 1912, all from *Corpus Christi Caller*.

17. Roth, "Texas Hurricane History," 5, 7, 31–32; "Tropical Storm Sweeps Close to Florida," August 15, 1915; "Work of Listing Dead and of Reconstruction Being Rushed," August, 20 1915, both from *Corpus Christi Caller*; additional death rate headlines are from *Corpus Christi Caller*, August 19 and 20, 1915. Two hundred seventy-five people died in that storm.

18. "Texas Weather Conditions Were Unusual Tuesday," August 18, 1915; "Wind Velocity Monday Running 38 MPH," August 17, 1915, both from *Corpus Christi Caller*; "Tex: Aransas Pass," *Manufacturers Record Baltimore: A Weekly Southern Industrial, Railroad and Financial Newspaper*, July 5, 1916.

19. "Beauteous Rains Bring Joy to South Texas," "Farmers Will Gather Bumper Crops in Fall," July 14, 1916, both from *Corpus Christi Caller*; Alfred J. Henry, "Forecasts and Warnings for August, 1916," *Monthly Weather Review*, August 1916, 461, accessed December 18, 2000, https://www.aoml.noaa.gov/hrd/hurdat/mwr_pdf/1916.pdf.

20. Henry, "Forecasts and Warnings."

21. Frank Clendening to Grace Clendening, August 19, 1017, in Harriett Johnson, *Hurricanes*, Kilgore Collection, #2357, Special Collections Room, Texas A&M University–Corpus Christi (hereafter cited as SCR); "Fifteen Deaths in

Storm Outside of Corpus Christi," *Corpus Christi Caller*, August 22, 1916; Henry, "Forecasts and Warnings."

22. "San Diego Feels Full Source of Wind," "Principal Waterfront Pleasure Resort Pier Practically Demolished by Tropical Storm," "Alice Devastated," "Cline's Flour Bluff Resort Carried Away by Wind and Storm," August 19, 1916; "Rockport Waterfront Swept Clear of Piers," August 22, 1916, all from *Corpus Christi Caller*; Henry, "Forecasts and Warnings."

23. Clendening letter; articles all from *Corpus Christi Caller*, August 19, 1916.

24. "In Spite of Losses, Citizens Feel Happy," *Corpus Christi Caller*, August 19, 1916; Henry, "Forecasts and Warnings"; "Alice Devastated," "San Diego Feels Full Force of Wind," "Bishop Is Hard Hit," August 19, 1916; "Fifteen Deaths in Storm Outside of Corpus Christi," August 22, 1916; "Renewed Agitation," August 26, 1916, all from *Corpus Christi Caller*.

25. Givens and Moloney, *Corpus Christi*, 207; W. B. Harmon to Simon Cohn, September 8, 1916, and Nick Kahl to Robert Mercer, September 12, 1916, both courtesy of PAPHA; Sears order card, September 29, 1916; Fields-Porter Lumber Company receipt, November 6, 1916, both courtesy of PAPHA.

26. "Tracks Damaged by Storm of Friday, August 16 Now Fully Restored," *Corpus Christi Caller*, August 31, 1916; Guthrie, "Aransas Harbor Terminal Railway."

27. "Port Aransas, Texas," in *War Department Annual Reports, 1919*, 1116.

28. "U. S. Will Help Dredge Bay at Corpus Christi," *Corpus Christi Caller*, September 11, 1919.

29. Roth, "Texas Hurricane History," 13; Dan Kilgore, "The 1919 Storm," *Nueces County Historical Commission Bulletin* 2 (November 1989): 51; John T. Carr, "Report 49: Hurricanes Affecting the Texas Gulf Coast" (Texas Water Development Board, June 1967), #3654 Kilgore Collection, SCR.

30. "Cotton Sells Off as Storm Passes," *New York Times*, September 13, 1919; "Gale Fails to Hit Coast," *San Antonio Express*, September 13, 1919.

31. Michael J. Ellis, *The Hurricane Almanac: 1986, Texas Edition* (Corpus Christi, TX: Caribbean Blue, 1986), 72; Theodore A. Fuller, *When the Century and I Were Young* (Sylva, NC: self-published, 1979), 244; Mike Cox, "Port Aransas 1919 Storm," TexasEscapes.com, June 24, 2015, accessed October 5, 2018, http:www.texasescapes.com/MikeCoxTexasTales/Port-Aransas-1919-Storm.htm; Givens and Moloney, *Corpus Christi*, 189; "Storm Drove Seagulls 400 Miles Inland to Ballinger," *Corpus Christi Caller*, September 24, 1919.

32. Russell P. Mozeney, unpublished manuscript, 9–10, collection 104, box 1.5, Russell P. Mozeney Papers, SCR.

33. Johnson, *Hurricanes*, 16; "Tropical Storm Expected to Sweep All of Gulf Coast," *San Antonio Express*, September 14, 1919.

34. Givens and Moloney, *Corpus Christi*, 190.

35. "Pleasure Party on Waldo Saved after Desperate Struggle," *Corpus Christi Caller*, September 23, 1919.

36. Johnson, *Hurricanes*, 17–18.

37. Allen and Taylor, "Aransas," 346, 369, 371.

38. Johnson, *Hurricanes*, 17.

39. Allen and Taylor, "Aransas," 348, 351, 354–57, 359.

40. Johnson, *Hurricanes*, 16; Ellis, *Hurricane Almanac*, 72.

41. "Woman Literally Tarred and Feathered by Storm," September 29, 1919; "Sister Thais Lost," September 24, 1919, both from *Corpus Christi Caller*; Georgia Nelson, "Hurricane of 1919 a Vivid Memory to Nun at Spohn," *Corpus Christi Caller-Times*, September 11, 1960; "Twenty Soldiers Lose Lives," September 17, 1919; "Deckman Praises Men Who Helped in Rescue Work," September 27, 1919, both from *Corpus Christi Caller*; Fuller, *When the Century and I*, 258–59.

42. Kilgore, "1919 Storm," 53–54; Ellis, *Hurricane Almanac*, 72.

43. "Detective Recalls Storm Washed 108 Dead to Ranch," *Corpus Christi Caller-Times*, January 18, 1959; "200 or More Dead, $10,000 Loss in Gulf Storm," *New York Times*, September 17, 1919; Britain R. Webb, "In Retrospection," personal memoir included in email to Jim Moloney from Mark Creighton, October 12, 2010; Fuller, *When the Century and I*, 260; "Find Body of Storm Victim on an Island," *Corpus Christi Caller*, February 12, 1920; J. R. Bluntzer, interview by J. L. Campbell, March 25, 1972, South Texas Archives, Texas A&M University–Kingsville; "Storm Victims Here Were 227," *Corpus Christi Caller*, March 31, 1920; Ellis, *Hurricane Almanac*, 73; Givens and Moloney, *Corpus Christi*, 194–95; Allen and Taylor, "Aransas," 372.

44. Alclair Mays Pleasant, interview by author, November 11, 2003, Corpus Christi, TX; Eleanor Dodson, interview by author, July 9, 2001, Corpus Christi, TX. For a more detailed account of the impact of the hurricane on Corpus Christi, see O'Rear, *Storm over the Bay*.

45. Dodson, interview; "Rehabilitation of Corpus Christi Is Assured by Nation through Red Cross," "Additional Troops Arrive in the City," September 23, 1919, both from *Corpus Christi Caller*; Roy Miller to W. P. Hobby, n.d., box 23, 1.2, #29; Chairman Relief Committee to Colonel M. R. Hilgard, October 4, 1919, box 23, 1.02, #43, both from Cities—Corpus Christi Records, LHR.

46. J. C. Houts to Gordon Boone, September 19, 1919, #17; J. G. Poindexter to Roy Miller, September 22, 1919, #5; Colonel W. D. Cope to Ralph Soape, September 27, 1919, #9, all from Cities—Corpus Christi Records, box 23, 1.02, LHR.

47. Cope to Soape; "Found Body of Storm Victim on an Island," *Corpus Christi Caller*, February 12, 1920; Johnson, *Hurricanes*, 16–18; *War Department Annual Reports, 1923*, 1006, 1008–9.

48. "Situation at Rockport and Other Towns Good," September 27, 1919; "Rockport Hard Hit," September 19, 1919, both from *Corpus Christi Caller*; Allen and Taylor, "Aransas," 364–68, 373.

49. Allen and Taylor, "Aransas," 364, 367–68, 373; "Protection Came First, and Then the Clean-up," *Corpus Christi Caller*, September 16, 1919.

50. "New Beach Better Than Previous One," *Corpus Christi Caller*, September 25, 1919; Ellis, *Hurricane Almanac*, 76; Johnson, *Hurricanes*, 18; Allen and Taylor, "Aransas," 374–75; J. R. Sprague, "How a Town Comes Back," *Saturday Evening Post*, June 19, 1920; "Credit Men Will Extend Help to Local Businesses," "Business Men Plan to Rebuild City," September 26, 1919, both from *Corpus Christi Caller*.

CHAPTER 16

1. J. H. Lang to Mayor, September 17, 1919, #3; Gordon Boone to Jack Alton, September 19, 1919, # 5; H. G. Thomas to Roy Miller, September 22, 1919, #7, all from Cities—Corpus Christi Records, C 23, box 1.04, LHR; "Report of $5,000,000 Fund Here Officially Denied," *Corpus Christi Caller*, September 24, 1919; Chairman Relief Committee to Colonel M. B. Hilgard, October 4, 1919, #15, C 23, Cities—Corpus Christi, box 1.04; "Army, State, and City Join with Civilians to Combat Disaster in Storm's Wake," *Corpus Christi Caller*, September 24, 1919; "Carlos Bee Will Introduce Bill for a Loan," *Corpus Christi Caller*, October 1, 1919.

2. "Carlos Bee Will Introduce Bill for a Loan," "Optimism Marks Meeting of All Kiwanians with Rotarians Here," October 3, 1919, both from *Corpus Christi Caller*.

3. By this time, the names Port Aransas and Harbor Island Basin had become synonymous, unless the town was referred to specifically. "Council Approves Breakwater Plan," *Corpus Christi Caller*, March 13, 1920; "Optimism Marks Meeting of All Kiwanians with Rotarians Here."

4. "Safe Harbor Plan Stands in River Bill," *Corpus Christi Caller*, March 28, 1920; General Executive Committee, *Corpus Christi Port Project*, October 20, 1921, Port of Corpus Christi Collection, section C, 2.05, 5, LHR; Corrie FitzSimmons, "Corpus Christi, 1919–1926" (unpublished manuscript), Tom Stewart papers; Roy Miller, "The Legislative History of the Port of Corpus Christi," December 17, 1926, Port of Corpus Christi Collection, section C, 2.02(a), 7, LHR.

5. H. R. Sutherland, "Petition for Establishment of a Navigation District," October 14, 1921, Port of Corpus Christi Collection, section C, 2.15, LHR; Mary Morrow, "A Brief History of the Port of Corpus Christi since 1926," Port of Corpus Christi Collection, section C, 2.02(b), 1; Harvey Weil, "The History of the Port of Corpus Christi: 1926–2001," Port of Corpus Christi, accessed July 19, 2006, https://portofcc.com/about/port/about-us/; Harry Plomarity, "History of the Port

of Corpus Christi," *Bulletin: Nueces County Historical Commission*, November 1997, 15–16; *Corpus Christi Port Project*, 3; "Letter from the Secretary of War," US Congress, document 321, 67th Cong., 2nd sess. (Washington, DC: Government Printing Office, 1922), 5.

6. Richard King, "A Brief Survey of the Activities in the Corpus Christi Area toward the Development of the Deep Water Port, 1919–1926" (unpublished manuscript), 2–3, Tom Stewart papers; Allen and Taylor, "Aransas," 378; Mary C. Riley, "The History of the Development of the Port of Corpus Christi" (master's thesis, University of Texas, 1951), 127, courtesy of Port Authority of Corpus Christi Archives, hereafter referred to as PACCA.

7. "Channel Being Dredged at Port Aransas by US," *Corpus Christi Caller*, December 23, 1919.

8. "Oil Traffic by Deep Water Coming Back," *Corpus Christi Caller*, June 25, 1920.

9. *War Department Annual Reports, 1923*, 1006.

10. "Letter from the Secretary of War," US Congress, document 321, 7.

11. The equivalent of $1,637,000 in 2016 currency is $21,949,244, and the equivalent of $1,150,000 is $15,419,445. See inflation calculator, Official Inflation Data, Alioth Finance, accessed April 21, 2019, http://www.in2013dollars.com/1921-dollars-in-2016?amount=1150000; "Letter from the Secretary of War," US Congress, document 321, 8, 9, 31.

12. The equivalent of $2,032,000 in 2016 currency is $27,245,488. See inflation calculator, Official Inflation Data, Alioth Finance, accessed April 21, 2019, http://www.in2013dollars.com/1921-dollars-in-2016?amount=1150000; "Letter from the Secretary of War," US Congress, document 321, 7.

13. The equivalent of $100,000 in 2016 currency is $1,340,822; of $90,000 is $1,206,740, and of $5,000,000 is $67,041,062. See inflation calculator, Official Inflation Data, Alioth Finance, accessed April 21, 2019, http://www.in2013dollars.com/1921-dollars-in-2016?amount=1150000; "Letter from the Secretary of War," US Congress, document 321, 10–11.

14. The equivalent of $900,000 in 2016 currency is $12,067,392, and of $2,000,000 is $26,816,425. See inflation calculator, Official Inflation Data, Alioth Finance, accessed April 22, 2019, http://www.in2013dollars.com/1921-dollars-in-2016?amount=1150000; Allen and Taylor, "Aransas," 378; "Letter from the Secretary of War," US Congress, document 321, 1–2, 10, 26, 27; 34, 38; Ben Johnson, "The Manchester Ship Canal," Historic UK, accessed December 11, 2019, https://www.historic-uk.com/HistoryMagazine/DestinationsUK/The-Manchester-Ship-Canal/; Sibley, *Port of Houston*, 158–62.

15. "Letter from the Secretary of War," US Congress, document 321, 34–36.

16. "Corpus Christi Gets Designation," May 25, 1922; "Corpus Christi Port Item Passes Senate," September 20, 1922; "Harding Signs Port Bill," September 23, 1922, all from *Corpus Christi Caller*.

17. The equivalent of $1,000,000 in 2016 currency is $14,286,131. See inflation calculator, Official Inflation Data, Alioth Finance, accessed April 22, 2019, http://www.in2013dollars.com/1922-dollars-in-2016?amount=1150000; Gilbert McGloin to Nueces County Commissioners, March 2, 1923, Port of Corpus Christi Collection, section C, 2.15, LHR; Roy Miller to Corpus Christi Port Development Association, December 29, 1922, Port of Corpus Christi Collection, section C, 203, LHR; Riley, "History of the Development," 145, 149, 151, courtesy of PACCA; Weil, "History of the Port"; Plomarity, "History of the Port," 16.

18. "New Gulf Harbor and Port at Corpus Christi, Texas," *Engineering New Record* 98 (January 1–June 30, 1927): 561; FitzSimmons, "Corpus Christi."

19. "New Gulf Harbor and Port," 562; FitzSimmons, "Corpus Christi"; Louise Wolffarth, "The Transportation Problems of Corpus Christi, Texas" (master's thesis, Texas Technological College, 1939), 85.

20. Outgoing trade, especially in cotton, however, was better. See "Record of Tonnage over Municipal Wharf for the Year 1915," Corpus Christi Government Collection, box 10, LHR; S. M. Wilcox, Corps of Engineers Map, *Survey for Safe and Adequate Harbor: Coast of Texas and Possible Ship Channels Leading to Rockport, Aransas Pass, and Corpus Christi. August, September, and October 1921*; "Letter from the Secretary of War," US Congress, document 321; L. M. Adams, "The Port of Corpus Christi," *Military Engineer* 24 (1932): 410.

21. Wolffarth, "Transportation Problems," 85; "Port Stands as Monument to Engineering Skills," September 4, 1926; "Ship Channel to Gulf Completed," July 18, 1926, both from *Corpus Christi Caller*.

CHAPTER 17

1. "Beaumont Is Enthralled by Intracoastal Canal Story as Told by Miller," *Beaumont Enterprise*, August 28, 1925; Roy Miller to Members of the Intracoastal Canal Association, January 25, 1927, courtesy of Gulf Intracoastal Canal Records, #663, Howard-Tilton Memorial Library, Tulane University; Wolffarth, "Transportation Problems," 117; Weil, "History of the Port," 8; Boyd-Campbell Company Manager to Robert Driscoll, November 30, 1927, Port of Corpus Christi Collection, section C, 2.03, LHR; Roy Miller, "The Legislative History of the Port of Corpus Christi," December 17, 1926, Port of Corpus Christi Collection, section C, 2.02(a), 8; *War Department Annual Reports, 1923*, 1006; Sargent and Bottin, *Case Histories*, 39; "Letter from the Secretary of War," US Congress, document 321, 4.

2. "Aransas Pass Oil Seekers Confident," May 28, 1924; "Sigmund After Casing to Test at Aransas Pass," January 6, 1925, both from *San Antonio Light*.

3. The equivalent of $4.50 in 2016 currency is $64.00, and of $25 is $350.90. See inflation calculator, Official Inflation Data, Alioth Finance, http://www.in2013dol lars.com/1928-dollars-in-2016?amount=25; Adams, "Port of Corpus Christi," 410; "Humble Brings Development to Port Aransas," *San Antonio Light*, August 28, 1928.

4. "Shipyard Warehouse at Rockport Burns," *San Antonio Express*, March 1, 1931; Allen and Taylor, "Aransas," 384; "Rockport Harbor—Gulf of Mexico, Rockport, TX USA," accessed April 25, 2019, Waymarking.com, http://www.waymarking.com; Shukalo, "Rockport, TX."

5. Kratz, "History of the Texas Shrimp Industry," 27, 29, 54–56; "Species Directory: White Shrimp," accessed May 30, 2019, National Oceanic and Atmospheric Administration Fisheries, http://www.fisheries.noaa.gov.

6. Montagna et al., *Characterization of Anthropogenic and Natural Disturbance*, 22; Kratz, "History of the Texas Shrimp Industry," 29, 33–36.

7. Kratz, "History of the Texas Shrimp Industry," 35–36, 41, 44, 54–56.

8. Kratz, "History of the Texas Shrimp Industry," 29–30, 52–54, 66; Montagna et al., *Characterization of Anthropogenic and Natural Disturbance*, 29–30.

9. "Boats Used on Fishing Trips," *Corpus Christi Times*, July 22, 1931; Charles R. Cable, interview by author, August 24, 2017, Carlsbad, NM.

10. Kevin Kokomoor, "In the Land of the Tarpon: The Silver King, Sport, and the Development of Southwest Florida, 1885–1916," *Journal of the Gilded Age and Progressive Era* 11, no. 2 (April 2012): 222; "Directory," in *Aransas Pass Texas: Where Sails Meet Rails*; Allen and Taylor, "Aransas," 385–86; Eisenhauer and Starnes, *Corpus Christi*, 21; Farley, *Fishing Yesterday's Gulf Coast*, 76; "S.A. Woman Leads Field of 25 in Tarpon Rodeo," 1932 First Deep Sea Roundup, courtesy of PAPHA; "The Granddaddy of Them All," Deep Sea Roundup, January 10, 2018, accessed July 10, 2019, https://www.facebook.com/deepsearoundup; "84th Annual Deep Sea Roundup, July 11–14, Finds New Ways to Support Education in the Texas Coastal Bend," Port Aransas and Mustang Island Tourism Bureau, accessed July 10, 2019, https://www.prnewswire.com/news-releases/84th-annual-deep-sea-roundup-july-11-14-finds-new-ways-to-support-education-on-the-texas-coastal-bend-300882702.html; J. Guthrie Ford, "The Aransas Pass Is King," courtesy of PAPHA; Ford and Creighton, *Images of America*, 56.

11. By this time, the Wood family, whose cattle business had been destroyed by the 1919 hurricane, had sold St. Joseph Island to Cyrus B. Lucas. It changed hands again and was finally purchased by Sid Richardson. See Alpha Kennedy Wood, *Texas Coastal Bend: People and Places* (San Antonio: Naylor, 1971), 133; Wes Ferguson, "There's Nothing but a Beach on St. Jo Island, but What More Do You

Need," *Texas Highways*, May 30, 2018, accessed July 12, 2019, https://texashighways.com/things-to-do/on-the-water/coast/there-s-nothing-but-a-beach-on-st-jo-island-what-more-do-you-need/; Farley, *Fishing Yesterday's Gulf Coast*, 3–4.

12. Farley, *Fishing Yesterday's Gulf Coast*, 8.

13. Farley, 9, 10.

14. Farley, 10–11.

CHAPTER 18

1. Ned C. Teller, interview by author, July 14, 2019, Corpus Christi, TX; "In Their Own Words: Hawsepiper," Project Liberty Ship: SS *John W. Brown*, accessed July 19, 2019, https://www.ssjohnbrown.org/blog/2015/2/7/coming-up-through-the-hawsepipe (if site unavailable, printout in author's possession). Because Aransas Pass pilots at this time were men, they will be referred to as such. Women are now active members of Aransas–Corpus Christi Pilots, or ARACOR. See "Port Introduces Their First Female Deputy Branch Pilot," KIII-TV, https://www.kiiitv.com; "Port Corpus Christi Pilot Board Set to Approve First Woman Deputy Port Pilot," Port of Corpus Christi, April 18, 2017, https://portofcc.com, both accessed July 29, 2019.

2. Wolffarth, "Transportation Problems," 91; "Gulf Export and Transport Co.," in *Record of American and Foreign Shipping* (New York: American Bureau of Shipping, 1923), 1242.

3. "Saunders, District Judge," *United States Circuit Courts of Appeals Reports* (Rochester, NY: Lawyers' Cooperative, 1904), 517; "Captain Edward T. Mercer Jr."; Edward Thomas Mercer, draft registration; B. J. Robbins memoirs, 3–4; "1853 and Before," both from Pilots Folder, PAPHA; "City Chosen as Distribution Point for Pure Oil Company," June 2, 1931; "Two Vessels Are in Port," April 6, 1931, both from *Corpus Christi Times*.

4. Teller, interview; Dan Parker, "Sea Is in Teller Blood," *Port Aransas South Jetty*, July 17, 2019; Cliff Russell, "Saga at Sea," *Corpus Christi Caller-Times*, November 10, 1957; Richard Watson, *Geologic Framework*, 11; Tunnell and Judd, *Laguna Madre*, 24.

5. White, *United States Early Radio History*; Proc, *Radio Communications*.

6. "Cotton Exports at Port May Pass 500,000 by Night," *Corpus Christi Times*, March 26, 1931; "Clyde Maritime Forum," Scottish Built Ships: The History of Shipbuilding in Scotland, accessed July 24, 2019, http://www.clydeships.co.uk/view.php?ref=8609; "Two Vessels Are in Port," *Corpus Christi Times*, April 6, 1931; "Netherlands Crew Rescued," in *Ships of the Esso Fleet in World War II* (New Jersey: Standard Oil, 1946), 66; Russell, "Saga at Sea"; Teller, interview.

7. Teller, interview; Parker, "Sea Is in Teller Blood."

8. "Letter from the Secretary of War," US Congress, document 321, 41; Neal Falgoust, "King Cotton History, Stadium Will Merge," *Corpus Christi Caller-Times*, January 11, 2004; Russell, "Saga at Sea"; "Ship Strikes Port Bascule Bridge Here," *Corpus Christi Times*, January 2, 1931; Murphy Givens, "Remember the Old Bascule Bridge?," *Corpus Christi Caller-Times*, April 12, 2000.

9. Teller, interview; Russell, "Saga at Sea."

10. Boyd-Campbell Company manager to Robert Driscoll, November 30, 1927, section C—Cities/Counties, #1, box 2.03, LHR; "Contract Let for Deepening of Channel," February 13, 1931, "Ship Channel Here to Be Deepened," August 31, 1933; "Corpus Christi Voted 32-Foot Port," August 21, 1935, all from *Corpus Christi Times*; Weil, "History of the Port."

CHAPTER 19

1. Montagna et al., *Characterization of Anthropogenic and Natural Disturbance*.

2. William A. White, Thomas A. Tremblay, Rachel L. Waldinger, and Thomas R. Calnan, *Status and Trends of Wetland and Aquatic Habitats of Texas Barrier Islands, Coastal Bend* (Austin: Texas General Land Office, 2006), fig. 47, accessed September 2, 2016, https://cbbep.org/publications/virtuallibrary/BarrierWetlandsCoastalBend.pdf.

3. Farley, *Fishing Yesterday's Gulf Coast*, 15.

4. "Whooping Crane," Encyclopedia.com, accessed July 30, 2019, https://www.encyclopedia.com/people/history/historians-miscellaneous-biographies/whooping-crane; Tunnell and Judd, *Laguna Madre* 98.

5. Montagna et al., *Characterization of Anthropogenic and Natural Disturbance*, 30, 32; Whitten, "Marine Biology"; Farley, *Fishing Yesterday's Gulf Coast*, 25.

6. Neal E. Armstrong and George H. Ward Jr., *Final Report: Point Source Loading Characterization of Corpus Christi Bay, Executive Summary* (Austin: University of Texas) accessed September 27, 2016, https://cbbep.org/publications/virtuallibrary/ccbnep30.pdf; Alan Peppard, "Islands of the Oil Kings: Part 1," *Dallas Morning News*, December 4, 2014. http://res.dallasnews.com/interactives/oilkings/part1/.

BIBLIOGRAPHY

ORAL INTERVIEWS

Cable, Charles R. Interview by author. August 24, 2017, Carlsbad, NM.
Dodson, Eleanor. Interview by author. July 9, 2001, Corpus Christi, TX.
Pleasant, Alclair Mays. Interview by author. November 11, 2003, Corpus Christi, TX.
Smith, Greg. Interview by author. March 5, 2018, Corpus Christi, TX.
Teller, Ned C. Interview by author. July 14, 2019, Corpus Christi, TX.

TRANSCRIBED INTERVIEWS

Bluntzer, J. R. Interview by J. L. Campbell. March 25, 1972. South Texas Archives, Texas A&M University–Kingsville.
Dunn, John. Interview. August 31, 1929, Corpus Christi, TX. Paul Schuster Taylor Papers, Bancroft Library, University of California, Berkeley.

PRIVATE COLLECTIONS

Author's personal papers.
Jim Moloney papers, Corpus Christi, TX.
Renato Ramirez documents, Corpus Christi, TX.
Tom Stewart papers, Corpus Christi, TX.

ARCHIVAL COLLECTIONS

Cities, City Government, and Port of Corpus Christi Collections. Local History Room, Corpus Christi Public Libraries.
Charts, maps, and documents. Port Aransas Preservation and Historical Association Museum.
Charts and maps. University of Texas Marine Science Institute.

Gulf Intracoastal Canal Records. Howard-Tilton Memorial Library, Tulane University, New Orleans, LA.

Kilgore Collection. Special Collections Room, Texas A&M University–Corpus Christi.

Paul Schuster Taylor Papers. Bancroft Library, University of California, Berkeley.

Port of Corpus Christi Authority Archives.

Russell P. Mozeney Papers. Special Collections Room, Texas A&M University–Corpus Christi.

ARTICLES, BOOKS, AND PRESENTATIONS
Primary Sources

Beaumont Enterprise, Bonham Daily Favorite, Brackett News Mail, Brownsville Daily Herald, Bryan Daily Eagle, Corpus Christi Caller, Corpus Christi Caller-Times, Corpus Christi Star, Corpus Christi Times, Daily Herald (San Antonio, TX), *Daily Picayune* (New Orleans, LA), *Daily San Antonio Light, Daily Victoria Advocate, El Paso Daily Herald, Engineering News Record, Flake's Daily Bulletin* (Galveston, TX), *Galveston Daily News, Galveston Tribune, Galveston Tri-Weekly News, Houston Chronicle, Laredo Weekly Times, New Orleans Daily Democrat, New York Herald, New York Times, Nueces Valley, Port Aransas Post, Port Aransas South Jetty, Ranchero* (Corpus Christi, TX), *Refugio County Press, Rockport Reporter and Messenger, San Antonio Express, San Antonio Herald, San Antonio Light, San Antonio Light and Gazette, Shiner Gazette, Victoria Advocate, Victoria Weekly Advocate, Weekly Corpus Christi Caller, Weekly Texas State Gazette* (Austin), *Wise County Messenger* (Decatur, TX)

Adams, L. M. "The Port of Corpus Christi." *Military Engineer* 24 (1932): 410–14.

"Almonte's Statistical Report on Texas." Translated by Carlos E. Castañeda. *Southwestern Historical Quarterly* 28 (1925): 177–221.

Bisso v. Inland Waterways Corporation. 50 US (1955). FindLaw. http://caselaw.findlaw.com/us-supreme-court/349/85.html.

"Captain Edward T. Mercer Jr." Find a Grave. https://www.findagrave.com/memorial/53334663/edward-thomas-mercer#clipboard.

Cline, Isaac M. "Special Report on the Galveston Hurricane of September 8, 1900." National Weather Service Heritage. https://vlab.noaa.gov/web/nws-heritage/-/galveston-storm-of-1900.

Committee on Publications, *Transactions of the American Society of Civil Engineers*. Vol. 54, part A. New York: American Society of Civil Engineers, 1905.

Bibliography

"Concrete Tanker Built Like Grain Elevator." *Popular Mechanics* 24 (November 1920).
Corpus Christi, Texas: June 1914. New York: Sanborn Map Company, 1914.
Crimmins, M. O., ed. "Notes and Documents: W. G. Freeman's Report on the Eighth Military Department." *Southwestern Historical Quarterly* 50 (1947): 350–57.
Dana, Napoleon Jackson Tecumseh. *Monterrey Is Ours! The Mexican War Letters of Lieutenant Dana, 1845–1847*. Edited by Robert H. Ferrell. Lexington: University Press of Kentucky, 1990.
Doubleday, Abner. *My Life in the Old Army*. Edited by Joseph Chance. Fort Worth: Texas Christian University Press, 1998.
Filby, P. William, ed. *Passenger and Immigration Lists Index, 1500s–1900s*. Farmington Hill, MI: Gale Research, 2012.
Ford, John Guthrie, ed. *The Mercer Logs: Pioneer Times on Mustang Island, Texas*. Port Aransas, TX: Port Aransas Preservation and Historical Association, 2012.
French, Samuel. *Two Wars: An Autobiography of General Samuel G. French*. Nashville: Confederate Veteran, 1901.
Frost, John. *Pictorial History of Mexico and the Mexican War*. Philadelphia: Thomas, Cowperthwait for J. A. Bill, 1849.
Fuller, Theodore A. *When the Century and I Were Young*. Sylva, NC: self-published, 1979.
General Taylor's Life, Battles and Correspondence. Philadelphia: T. C. Clarke, 1847. https://archive.org/stream/brilliantnationaoophil/brilliantnationaoophil_djvu.t.
Grant, Ulysses S. *Personal Memoirs, 1885–1886*. Chap. 4. http://www.bartleby.com/1011/4.html.
"Gulf Export and Transport Co." In *Record of American and Foreign Shipping*. New York: American Bureau of Shipping, 1923.
Heath, C. C., and William R. Roberts. "Code of Signals" flyer. August 1, 1874.
Helm, Mary S. *Scraps of Early Texas History*. Austin: Eakin Press, 1987.
Henry, Alfred J. "Forecasts and Warnings for August, 1916." *Monthly Weather Review*, August 1916. https://www.aoml.noaa.gov/hrd/hurdat/mwr_pdf/1916.pdf.
Henry, W. S. *Campaign Sketches of the War with Mexico*. New York: Harper Brothers, 1847.
Hitchcock, Ethan Allen. *Fifty Years in Camp and Field: Diary of Major-General Ethan Allen Hitchcock, USA*. Edited by W. A. Croffut. New York: G. P. Putnam's Sons, 1909.
Improvement of Aransas Pass and Bay up to Rockport and Corpus Christi. Report of Major Ernst, Chief of Engineers, US Army. June 1887. In *Chief of Engineers United States Army, to the Secretary of War for the Year 1888, in Four Parts*. Part 2. Washington, DC: Government Printing Office, 1888.
Kent, William. "Ropes Pass, Texas." *Engineering Magazine* 3 (June 1892): 340–45.

"Letter from the Secretary of War." US Congress. House Documents. 62nd Cong., 3rd sess. December 2, 1912–March 4, 1913. Washington, DC: Government Printing Office, 1913.

"Letter from the Secretary of War." US Congress, document 321. 67th Cong., 2nd sess. Washington, DC: Government Printing Office, 1922.

Lewis V. Haupt v. United States in Cases Argued and Decided in the Supreme Court of the United States, October Term, 1920. Rochester, NY: Lawyers' Cooperative, 1922.

Mayo, John J. *The British Code List for 1874 for the Use of Ships at Sea, and for Signal Stations.* London: Sir William Mitchell, 1874.

Meade, George. *The Life and Letters of George Gordon Meade, Major General, United States Army.* Vol. 1. New York: Charles Scribner's Sons, 1913.

"New Gulf Harbor and Port at Corpus Christi, Texas." *Engineering New Record* 98 (January 1–June 30, 1927): 560–62.

Official Records of the Union and Confederate Navies in the War of the Rebellion. Series 1, vol. 19. Washington, DC: Government Printing Office, 1905. https://texashistory.unt.edu/ark:/67531/metapth192854/?q=Sabine%20Pass.

Proposed Bombing and Machine Gun Restricted Areas along Gulf of Mexico from San Luis Pass to Aransas Pass: Index Map. Galveston, TX: Engineer District, War Department, January 1941.

Records of the Bureau of the Census. National Archives, Washington, DC.

"Report of the Chief of Engineers, US Army." *Annual Reports of the War Department for the Fiscal Year Ended June 30, 1901.* Parts 2 and 3. Washington, DC: Government Printing Office, 1901.

"Report of the Chief of Engineers, US Army." *Annual Reports of the War Department.* Washington, DC: Government Printing Office, 1904.

"Robert Mercer." Lancashire, England, Church of England Marriages and Banns, 1754–1936. Ancestry.com, 2012.

"Saunders, District Judge." *United States Circuit Courts of Appeals Reports.* Rochester, NY: Lawyers' Cooperative, 1904.

Shorey, Henry A. *The Story of the Maine Fifteenth, Being a Brief Narrative of the More Important Events in the History of the Fifteenth Maine Regiment.* Bridgton, ME: Press of the *Bridgton News*, 1860.

"Ten Concrete Ships Now Building at Port Aransas." *Concrete and Engineering News* 31 (January 1919).

"Tex: Aransas Pass." *Manufacturers Record Baltimore: A Weekly Southern Industrial, Railroad and Financial Newspaper,* July 5, 1916.

United States Coast Pilot: Atlantic Coast. Part 7, *Gulf of Mexico from Key West to the Rio Grande.* Washington, DC: US Treasury Department, US Coast and Geodetic Survey Office, 1896.

von Blücher, Maria. *Maria von Blücher's Corpus Christi: Letters from the South Texas Frontier, 1849–1879*. Edited by Bruce S. Cheeseman. College Station: Texas A&M University Press, 2002.

Wagner, Frank, ed. *Bérenger's Discovery of Aransas Pass: A Translation of Jean Bérenger's French Manuscript by William M. Carroll*. Corpus Christi, TX: Friends of the Corpus Christi Museum, 1983.

War Department Annual Reports, 1919. Vol. 2, *Report of the Chief of Engineers*. Washington, DC: Government Printing Office, 1919.

War Department Annual Reports, 1923. *Report of the Chief of Engineers*. Washington, DC: Government Printing Office, 1923.

The War of the Rebellion: A Compilation of the Official Records of the Union and Confederate Armies. Series 1, vol. 9. Published by an act of Congress, June 16, 1880.

Water Terminal and Transfer Facilities: Letter from the Secretary of War. Washington, DC: Government Printing Office, 1921.

Whiting, Daniel P. *A Soldier's Life: Memoirs of a Veteran of 30 Years of Soldiering, Seminole Warfare in Florida, the Mexican War, Mormon Territory, and the West*. Edited by Murphy Givens. Corpus Christi, TX: Nueces Press, 2011.

Secondary Sources

"19th Century Steamships." Bureau of Ocean Energy Management. https://www.boem.gov/environment/19th-century-steamships.

"84th Annual Deep Sea Roundup, July 11–14, Finds New Ways to Support Education in the Texas Coastal Bend." Port Aransas and Mustang Island Tourism Bureau. https://www.prnewswire.com/news-releases/84th-annual-deep-sea-roundup-july-11-14-finds-new-ways-to-support-education-on-the-texas-coastal-bend-300882702.html.

"An Alchemists Glossary of Terms, Definitions, Formulas & Concoctions - Part 2." The Third Millennium. http://www.3rd1000.com/alchemy/alchemyterms2.htm#N.

Allen, William, and Sue Hastings Taylor. "Aransas: The Life of a Texas Coastal County." Vol. 2. Unpublished manuscript. Aransas County Historical Society, 1997.

Allhands, J. L. "Lott, Uriah." *Handbook of Texas Online*. https://www.tshaonline.org/handbook/entries/lott-uriah.

Alperin, Lynn. *History of the Gulf Intracoastal Waterway*. Navigation History NWS-83–9, January 18, 1983.

"Aransas Pass, Texas." In *The Handbook of Texas*. Edited by Walter Prescott Webb. Austin: Texas State Historical Association, 1952.

Aransas Pass Texas: Where Sails Meet Rails. Aransas Pass: Chamber of Commerce, August 1913. www.larryray.com/aransas-pass-tx.html.

Armstrong, Neal E., and George H. Ward Jr. *Final Report: Point Source Loading Characterization of Corpus Christi Bay, Executive Summary*. Austin: University of Texas. https://cbbep.org/publications/virtuallibrary/ccbnep30.pdf.

"Artillery." US Army Ordnance Corps. http://www.goordnance.army.mil/history/Staff%20Ride/STAND%203%20ARTILLERY%20AND%20SMALL%20ARMS/ARTILLERY%20IN%20THE%20CIVIL%20WAR.pdf.

Artsy, Avishay. "How the Jewfish Got Its Name." Jewish Telegraphic Agency. https://www.jta.org.

Atwater, Gordon A. "Navigation." *Collier's Encyclopedia*. USA: Crowell-Collier Education Corp., 1969.

Barragy, T. G. *Gathering Texas Gold: Frank Dobie and the Men Who Saved the Longhorns*. Corpus Christi, TX: Cayo del Grullo Press, 2003.

"Barrier Island Interior Wetlands." Texas A&M AgriLife Extension. http://texaswetlands.org/wetland-types/barrier-island-interior-wetlands.

"The Battle of Corpus Christi." The American Civil War. https://www.mycivilwar.com/battles/620816.html.

Baughman, James P. *Charles Morgan and the Development of Southern Transportation*. Nashville: Vanderbilt University Press, 1968.

———. "The Evolution of Rail-Water Systems of Transportation in the Gulf Southwest, 1836–1890." *Journal of Southern History* 34 (1968): 357–81.

Beemer, Rod. *The Deadliest Woman in the West: Mother Nature on the Prairies and Plains, 1800–1900*. Caldwell, ID: Caxton Press, 2006.

Bixel, Patricia Bellis, and Elizabeth Hayes Turner. *Galveston and the 1900 Storm: Catastrophe and Catalyst*. Austin: University of Texas Press, 2000.

Bluntzer, John Lloyd. "The Texas-Mexican Railroad." Presentation at the South Texas Historical Association, Rockport, TX, April 6, 2019.

Briscoe, Eugenia Reynolds. *City by the Sea: A History of Corpus Christi*. New York: Vantage Press, 1985.

Brown, L. F., and J. L. Brewton. *Environmental Geologic Atlas of the Texas Coastal Zone: Corpus Christi Area*. Austin: University of Texas, Bureau of Economic Geology, 1976.

Bruseth, James E., and Toni S. Turner. *From a Watery Grave: The Discovery and Excavation of La Salle's Shipwreck*, La Belle. College Station: Texas A&M University Press, 2005.

"Camp Corpus Christi." *Handbook of Texas Online*. http://www.tshaonline.org/handbook/online/articles/qbc10.

Canales, Herb. "¡Viva el Rey Alonso! The Legend of Who Discovered and Named Corpus Christi Bay." *Journal of South Texas* 24 (2011): 54–75.

Carnes, Mark, and John Garraty. *Mapping America's Past*. New York: Henry Holt, 1996.
Carr, William R. "Some Plants of the South Texas Sand Sheet." http://w3.biosci.utexas.edu/prc/DigFlora/WRC/Carr-SandSheet.html.
Cartwright, Gary. *Galveston: A History of the Island*. Fort Worth: Texas Christian University Press, 1991.
Centennial History of Corpus Christi. Corpus Christi, TX: *Corpus Christi Caller-Times*, 1952.
Chakrabarty, Amitava. "What Marine Communication Systems Are Used in the Maritime Industry?" Marine Insight. https://www.marineinsight.com/marine-navigation/marine-communication-systems-used-in-the-maritime-industry/.
Cheeseman, Bruce S. "Richard King." *Handbook of Texas Online*. https://www.tshaonline.org/handbook/entries/king-richard.
Chipman, Donald E. *Spanish Texas, 1519–1821*. Austin: University of Texas Press, 1992.
Chipman, Donald E., and Robert S. Weddle. "How Historical Myths Are Born . . . and Why They Seldom Die." *Southwestern Historical Quarterly* (January 2013): 227–60.
"Civil War Weapons." History.net. https://www.historynet.com/civil-war-weapons.
"Clyde Maritime Forum." Scottish Built Ships: The History of Shipbuilding in Scotland. http://www.clydeships.co.uk/view.php?ref=8609.
"Coastal Texas II." Texas Historical Sites. http://www.northamericanforts.com/West/tx-coast2.html.
Coffee, Phyllis. "Logs Reveal Texas Gulf Coast History, 1866–1900." *Southwestern Historical Quarterly* 62 (1959): 227–32.
Cooper, George. "The Railroads of South Texas." Presentation at the South Texas Historical Association, Rockport, TX, April 6, 2019.
Cox, Mike. "Port Aransas 1919 Storm." TexasEscapes.com. June 24, 2015. http:www.texasescapes.com/MikeCoxTexasTales/Port-Aransas-1919-Storm.htm.
Crnkovich, Kellie, comp. "Robert Ainsworth Mercer." http://www.rootsweb.ancestry.com/txaransa/mercer.htm.
Cunningham, Atlee M. *Corpus Christi Water Supply Documented History: 1852–1997*. Corpus Christi, 1997.
Current, Richard N. "U. S. Civil War," *Collier's Encyclopedia*. Vol. 6. Edited by William D. Halsey. USA: Crowell-Collier Educational Corp., 1969.
Davis, Jack E. *The Gulf: The Making of an American Sea*. New York: Liveright, 2017.
Delaney, Norman C. "Hanging Dedication." Speech delivered at the Texas State Historical Marker Dedication, Corpus Christi, TX, May 4, 2017.

———. *The Maltby Brothers' Civil War*. College Station: Texas A&M University Press, 2013.

———. "Two Civil War Hangings in Corpus Christi. " *Nueces County Historical Commission Bulletin* 6 (March 2016).

———. "Whiskey and the Battle of Corpus Christi." The Texas Story Project, Bullock Texas State History Museum. https://www.thestoryoftexas.com/discover/texas-story-project/whiskey-shells-corpus-christi.

Docevski, Boban. "Depth Sounding Techniques That Preceded the Modern Day SONAR Technology." *Vintage News*, February 23, 2017. https://www.thevintage news.com/2017/02/23/depth-sounding-techniques-that-preceded-the-modern-day-sonar-technology/.

Doughty, Robin W. "Sea Turtles in Texas: A Forgotten Commerce." *Southwestern Historical Quarterly* 88 (July 1984): 43–70.

Downey, Fairfax. *Texas and the War with Mexico*. New York: American Heritage, 1961.

"Dugout Canoes." Cherokee Heritage Center. http://www.cherokeeheritage.org/attractions/dugout-canoe/.

Dunn, Lawrence. *The World's Tankers*. London: Adlard Coles, 1956.

Eisenhauer, Anita, and Gigi Starnes. *Corpus Christi, Texas: A Picture Postcard History*. Corpus Christi, TX: Anita's Antiques, 1987.

Ellis, Michael J. *The Hurricane Almanac: 1986, Texas Edition*. Corpus Christi, TX: Caribbean Blue, 1986.

"Facts about Sulfur." Live Science. http://www.livescience.com/28939-sulfur.html.

Farley, Barney. *Fishing Yesterday's Gulf Coast*. College Station: Texas A&M University Press, 2002.

Felgar, Robert Patterson. "Texas in the War for Southern Independence." Master's thesis, University of Texas, 1935.

Ferguson, Wes. "There's Nothing but a Beach on St. Jo Island, but What More Do You Need." *Texas Highways*, May 30, 2018. https://texashighways.com/things-to-do/on-the-water/coast/there-s-nothing-but-a-beach-on-st-jo-island-what-more-do-you-need/.

"Field Artillery in the Civil War." cwartillery.com. https://cwartillery.com/FA/FA.html.

FitzSimmons, Corrie. "Corpus Christi, 1919–1926." Unpublished manuscript. Tom Stewart papers, Corpus Christi, TX.

Ford, J. Guthrie. "Fort Semmes." *Handbook of Texas Online*. http://www.tshaonline.org/handbook/online/articles/qcf29.

———. *A Texas Island*. Port Aransas, TX: USA Hurrah Publishing, 2008.

Ford, J. Guthrie, and Mark Creighton. *Images of America: Port Aransas*. Charleston, SC: Arcadia Publishing, 2010.

———. "Our Bali Ha'i and a Watery Railroad." *PAPHA Newsletter*, March 2010.

Foster, William C., ed. *The La Salle Expedition to Texas: The Journal of Henri Joutel, 1684–1687*. Austin: Texas State Historical Association, 1998.

Francaviglia, Richard V. *From Sail to Steam: Four Centuries of Texas Maritime History: 1500–1900*. Austin: University of Texas Press, 1998.

Frank, Norman. "Shipyards in Rockport." Historical Marker Database. https://hmdb.org/marker.asp?marker=58824.

Frankl, Guido, and Ramon N. Garcia. *Soil Survey: Nueces County, Texas*. Series 1960, no. 26. US Department of Agriculture Soil Conservation Service, 1965.

Freese, Simon, and Deborah Lightfood Sizemore. *A Century in the Works: Freese and Nichols Consulting Engineers*. College Station: Texas A&M University Press, 1994.

Gillett, Mary C. *The Army Medical Department, 1818–1865*. Washington, DC: Government Printing Office, 1987.

Givens, Murphy. *Great Tales from the History of South Texas*. Corpus Christi, TX: Nueces Press, 2012.

Givens, Murphy, and Jim Moloney. *Corpus Christi: A History*. Corpus Christi, TX: Nueces Press, 2011.

"Goliath Grouper: Fish." *Encyclopedia Britannica*. https://www.britannica.com.

Graf, Leroy P. "Colonizing Projects in Texas South of the Nueces, 1829–1845." *Southwestern Historical Quarterly* 50 (April 1947): 431–38.

"The Granddaddy of Them All." Deep Sea Roundup, January 10, 2018. https://www.facebook.com/deepsearoundup.

Guckian, William J., and Ramon N. Garcia. *Soil Survey of San Patricio and Aransas Counties, Texas*. US Department of Agriculture and Texas Agriculture Experiment Station, July 1979.

Guthrie, Keith. "Aransas Harbor Terminal Railway." *Handbook of Texas Online*. https://www.tshaonline.org/handbook/entries/aransas-harbor-terminal-railway.

———. "Aransas Pass, TX." *Handbook of Texas Online*. http://www.tshaonline.org/handbook/online/articles/hfa06.

Hall, Martin Hardwick. *Sibley's New Mexico Campaign*. Albuquerque: University of New Mexico Press, 1960.

Harwood, Miller, and W. A. Serivner. *Fabulous Port Aransas*. Travel South Texas. 1949. www.stxmaps.com/go/fabulous-port-aransas3.html.

Haugh, George F. "Notes and Documents: History of the Texas Navy." *Southwestern Historical Quarterly* 58 (April 1960): 572–78.

"Hereford." The Cattle Site. www.thecattlesite.com/breeds/beef/14/hereford.

"Historic New Albany." www.historicnewalbany.com.

"History of Dredging." Start Dredging.com. https://www.startdredging.com.

"History of Navigation at Sea," Water Encyclopedia. http://www.waterencyclopedia.com/Mi-Oc/Navigation-at-Sea-History-of.html.

"History of the Astrolabe." Astrolabe. www.astrolabe.org.

"How Do Hurricanes Form?" National Ocean Service, NOAA. https://oceanservice.noaa.gov.

"Information about Sea Turtles: Green Sea Turtles." Sea Turtle Conservancy. https://conserveturtles.org.

Johnson, Ben. "The Manchester Ship Canal." Historic UK. https://www.historicuk.com/HistoryMagazine/DestinationsUK/The-Manchester-Ship-Canal/.

Jones, Fred. *Flora of the Texas Coastal Bend*. Sinton, TX: Rob and Bessie Welder Wildlife Foundation, 1982.

Keegan, John. *The American Civil War: A Military History*. New York: Alfred A. Knopf, 2009.

Kenmotsu, Nancy, and Susan Dial. "Native Peoples of the Coastal Prairies and Marshlands in Early Historic Times." Texas beyond History. http://www.texasbeyondhistory.net/coast/peoples/.

Kennedy, Robin Borglum. *Mary's Story: Mary Borglum's Story from the Mountains of Anatolia to the Mountains of South Dakota*. North Charleston, SC: CreateSpace, 2013.

Kilgore, Dan. "The 1919 Storm." *Nueces County Historical Commission Bulletin* 2 (November 1989).

King, Richard. "A Brief Survey of the Activities in the Corpus Christi Area toward the Development of the Deep Water Port, 1919–1926." Unpublished manuscript. Tom Stewart papers, Corpus Christi, TX.

King Ranch: 100 Years of Ranching, 1853–1953. Corpus Christi, TX: *Corpus Christi Caller-Times*, 1953.

Kinnan, Bob. "Echoes in the Dust: Captain Richard King and Rancho de Santa Gertrudis, 1865–1885." Presentation at the East Texas Historical Association, Galveston, TX, October 13, 2017.

Kirchner, Paul G., and Clayton L. Diamond. "Unique Institutions, Indispensable Cogs, and Hoary Figures: Understanding Pilotage Regulation in the United States." *University of San Francisco School of Law Maritime Law Journal* 23, no. 1 (2010–2011). http://www.americanpilots.org/document_center/Activities/Unique_Institutions__Indispensable_Cogs__and_Hoary_Figures_Understanding_Pilotage_Regulation_in_the_United_States.pdf.

Kohout, Martin Donell. "Duval County." *Handbook of Texas Online*. http://www.tshaonline.org/handbook/online/articles/hcd11.

Kokomoor, Kevin. "In the Land of the Tarpon: The Silver King, Sport, and the Development of Southwest Florida, 1885–1916." *Journal of the Gilded Age and Progressive Era* 11, no. 2 (April 2012): 191–224.

Kolker, Claudia. "The Salty Lagoon." *Texas Parks and Wildlife*, July 2003.
Kratz, Jeremiah Frederick. "A History of the Texas Shrimp Industry." Master's thesis, University of Texas, 1963.
Lea, Tom. *The King Ranch*. Vol. 1. Boston: Little, Brown, 1957.
Leatherwood, Art. "Corpus Christi Bay." *Handbook of Texas Online*. http://www.tshaonline.org/handbook/onlie/articles/rrc05.
Lessoff, Alan. *Where Texas Meets the Sea: Corpus Christi and Its History*. Austin: University of Texas Press, 2015.
Lipscomb, Carol A. "Karankawa Indians." *Handbook of Texas Online*. http://www.tshaonline.org/handbook/online/articles/bmk05.
Macdonald, Jessica. "The Damage of Dynamite Fishing." Scuba Diver Life. September 1, 2014. https://scubadiverlife.com/damage-dynamite-fishing.
Malsch, Brownson. *Indianola: The Mother of West Texas*. Austin: State House Press, 1988.
McComb, David G. *Galveston: A History*. Austin: University of Texas Press, 1986.
Moneyhon, Carl. *Edmund J. Davis of Texas: Civil War Leader, Reconstruction Governor*. Fort Worth: Texas Christian University Press, 2010.
Montagna, Paul A., Scott Holt, Christine Ritter, Sharon Herzka, and Kenneth H. Dunton. *Characterization of Anthropogenic and Natural Disturbance on Vegetated and Unvegetated Bay Bottom Habitats in the Corpus Christi Bay National Estuary Program Study Area*. Port Aransas: University of Texas at Austin Marine Science Institute, May 1998. www.//cbbep.org/publications/virtuallibrary/cc25a.pdf.
"Mustang Island State Park." Texas Parks and Wildlife Department. https://tpwd.texas.gov/state-parks/mustang-island/nature.
National Weather Service Heritage. "NWS Timeline." https://vlab.noaa.gov/web/nws-heritage/explore-nws-history#event-debut-of-the-daily-weather-map.
"Navigation in the 18th Century." Penobscot Marine Museum. https://penobscotmarinemuseum.org/pbho-1/history-of-navigation/navigation-18th-century.
"Navigation in the 19th to 20th Centuries." Penobscot Marine Museum. https://penobscotmarinemuseum.org/pbho-1/history-of-navigation/navigation-19th-20th-centuries.
"Navigation of the American Explorers, 15th to 17th Centuries." Penobscot Marine Museum. http://www.penobscotmarinemuseum.org/pbho-1/history-of-navigation/navigation-american-explorers-15th-17th-centuries.
Newcombe, W. W., Jr. *The Indians of Texas: From Prehistoric to Modern Times*. Austin: University of Texas Press, 1961.
Nixon, Jay. *Stewards of a Vision: A History of King Ranch*. Hong Kong: King Ranch, 1986.
Nummedal, Dag, ed. *Sedimentary Processes and Environments along the Louisiana-Texas Coast*. Baton Rouge, LA: Geological Society of America, 1982.

O'Rear, Mary Jo. *Storm over the Bay: The People of Corpus Christi and Their Port*. College Station: Texas A&M University Press, 2009.

"Oyster Life Cycle." University of Maryland Center for Environmental Science. https://hatchery.hpl.umces.edu/oyster-life-cycle/.

"Padre Island National Seashore: The Civil War." US National Park Service. https://www.nps.gov/pais/learn/historyculture/civil-war.htm.

"Padre Island National Seashore: Dunn Ranch." National Park Service. https://www.nps.gov/pais/learn/historyculture/dunn-ranch.htm.

Payne, Darwin. "Camp Life in the Army of Occupation: Corpus Christi, July 1845–March 1846." *Southwestern Historical Quarterly* 73 (January 1970): 326–42.

Pede, Charles N. "Discipline Rather Than Justice: Courts-Martial and the Army of Occupation at Corpus Christi, 1845–1846." *Army History*, Fall 2016, 35–50.

Peppard, Alan. "Islands of the Oil Kings: Part 1." *Dallas Morning News*, December 4, 2014. http://res.dallasnews.com/interactives/oilkings/part1/.

Pierce, Frank C. *A Brief History of the Lower Rio Grande Valley*. Menosha, WI: George Banta, 1917.

Pilkey, Orrin H. *A Celebration of the World's Barrier Islands*. New York: Columbia University Press, 2003.

"Pilot Flags." Flags of the World. http://www.crwflags.com/fotw/flags/xf-pilt.html.

Plomarity, Harry. "History of the Port of Corpus Christi." *Bulletin*: *Nueces County Historical Commission*, November 1997.

"Port Corpus Christi Pilot Board Set to Approve First Woman Deputy Port Pilot." Port of Corpus Christi. April 18, 2017. https://portofcc.com.

"Port Introduces Their First Female Deputy Branch Pilot." KIII-TV. https://www.kiiitv.com.

Poskett, James. "Edward Massey." *Longitude Essays*. Cambridge Digital Library. https://cudl.lib.cam.ac.uk/view/ES-LON-00028/1.

Price, W. Armstrong. "Reduction of Maintenance by Proper Orientation of Ship Channels through Tidal Inlets." Presentation at Second Conference of Coastal Engineers, Houston, 1951. Published 1952.

Proc, Jerry. "A Brief History of Naval Radio Communications." http://jproc.ca/rrp/nro_his.html.

"Red Drum (*Sciaenops ocellatus*)." Texas Parks and Wildlife. https://tpwd.texas.gov/huntwild/wild/species/reddrum/.

"Report 49: Hurricanes Affecting the Texas Gulf Coast." Texas Water Development Board. June 1967.

Reséndez, Andrés. *A Land So Strange*: *The Epic Journey of Cabeza de Vaca*. New York: Basic Books, 2007.

Ricklis, Robert A. *The Karankawa Indians of Texas*: *An Ecological Study of Cultural Tradition and Change*. Austin: University of Texas Press, 1996.

———. "Prehistoric and Early Historic People and Environment in the Corpus Christi Bay Area." Coastal Bend Bays and Estuaries Project. http://www.cbbep.org/publications/virtuallibrary/ricklis.html.

———. *The Prehistory of the Texas Coastal Zone: 10,000 Years of Changing Environment and Culture*. Texas beyond History. http://www.texasbeyondhistory.net/coast/prehistory/images/intro.html.

Riley, Mary C. "The History of the Development of the Port of Corpus Christi." Master's thesis, University of Texas, 1951.

"Rockport Harbor—Gulf of Mexico, Rockport, TX USA." http://www.waymarking.com.

Roth, David. "Texas Hurricane History." Unpublished manuscript. National Weather Service, 2000.

Sargent, Francis E., and Robert R. Bottin Jr. *Case Histories of Corps Breakwater and Jetty Structures*. Washington, DC: Department of the Army, US Corps of Engineers, 1989.

"Screwmen, Spidermen, and Cotton's Gilded-Age Gargantua." The History Bandits. https://thehistorybandits.com/2015/02/13/screwmen-spidermen-and-cottons-gilded-age-gargantua/.

"Sea Turtles: Adaptations." SeaWorld. https://seaworld.org/animals/all-about/sea-turtles/adaptations/.

Sheire, James W. *Padre Island National Seashore Historic Resource Study*. Washington, DC: US Department of the Interior National Park Service, Office of History and Historic Architecture, August 1971.

Ships of the Esso Fleet in World War II. New Jersey: Standard Oil, 1946.

Shook, Phil H. "Farley Boats and Tarpon: The Farley Family Boatbuilders." *Texas Parks and Wildlife*, October 1995.

Shukalo, Alice M. "Rockport, TX." *Handbook of Texas Online*. http://www.tshaonline.org/handbook/online/articles/hgr05.

Sibley, Mary McAdams. *The Port of Houston: A History*. Austin: University of Texas Press, 1968.

Skaggs, Jimmy M. "Cattle Trailing." *Handbook of Texas Online*. http://www.tshaonline.org/handbook/online/articles/ayc01.

Slattery, Margaret Patrice. *Promises to Keep: A History of the Sisters of Charity of the Incarnate Word*. Vol. 2. San Antonio: Sisters of Charity of the Incarnate Word, 1995.

Smith, Greg. "The Dunn Family in Nueces County and Ranching on Padre Island." Presentation at the Nueces County Historical Society, Corpus Christi, TX, April 1, 2014.

———. "Ranching on the Islands." Presentation at the South Texas Historical Association, Port Aransas, TX, November 2, 2013.

Soil Survey of Padre Island National Seashore, Texas, Special Report. US Department of Agriculture, Natural Resources Conservation Service, and US Department the Interior, National Park Service, 2005.

"Species Directory: White Shrimp." National Oceanic and Atmospheric Administration Fisheries. http://www.fisheries.noaa.gov.

Sprague, J. R. "How a Town Comes Back." *Saturday Evening Post*, June 19, 1920.

Spurlin, Charles D. *Texas Volunteers in the Mexican War*. Austin, TX: Eakin Press, 1998.

"Stadimeter." Smithsonian National Museum of American History. https://amhistory.si.edu/navigation/type.cfm?typeid=13.

"Steam Dredges." *Steam Shovel and Dredge* 13 (November 1909).

Stephens, A. Ray, and William K. Holmes. *Historical Atlas of Texas*. Norman: University of Oklahoma Press, 1989.

Stewart, Tom W. "History of the Aransas Pass Jetties." Presentation at the Texas Branch of the American Society of Civil Engineers, Corpus Christi, TX, March 22, 2013.

Stranahan, Pam. "Why Was WWI Called 'The Great War?'" History Center for Aransas County. https://www.thehistorycenterforaransascounty.org/history-mystery-1/why-was-wwi-called-%E2%80%98the-great-war%E2%80%99%3F.

Swinehart, James B. "Geology." *Encyclopedia of the Great Plains*. http://plainshumanities.unl.edu/encyclopedia/doc/egp.pe.028.

Tan, Henry. "Underwater Explosion." University of Aberdeen. 2008. https://homepages.abdn.ac.uk/h.tan/pages/teaching/explosion-engineering/Underwater-I.pdf.

"Tarpon Facts." TarponFish.com. https://www.tarponfish.com.

Terrell, J. L., and James A. Cook. "Magnolia Petroleum Company." *Handbook of Texas Online*. http://www.tshaonline.org/handbook/online/articles/dom01.

Texas Historic Sites Atlas. https://atlas.thc.state.tx.us/details/5355001519.

Thompson, Jerry. *Cortina: Defending the Mexican Name in Texas*. College Station: Texas A&M University Press, 2007.

Thompson, John T. "Governmental Responses to the Challenges of Water Resources in Texas." *Southwestern Historical Quarterly* 70 (1966): 44–64.

Thorton, R. H. "Taylor's Trail in Texas." *Southwestern Historical Quarterly* 70 (July 1977): 7–21.

"Treaties of Velasco." *Handbook of Texas Online*. http://www.tshaonline.org/handbook/online/articles/mgt05.

Triplett, Henry F., and Ferdinand A. Hauslein. *Civics: Texas and Federal*. Houston: Rein and Sons, 1912.

Tunnell, John W., Jr., Jean Andrews, Joe C. Barrera, and Fabio Moretzsohn. *Encyclopedia of Texas Seashells: Identification, Ecology, Distribution, and History*. College Station: Texas A&M University Press, 2010.

Tunnell, John W., Jr., and Frank W. Judd. *The Laguna Madre of Texas and Tamaulipas*. College Station: Texas A&M University Press, 2002.
US World War I Draft Registrations. Ancestry. https://search.ancestry.com/search.
Varnum, Robin. *Álvar Núñez Cabeza de Vaca: American Trailblazer*. Norman: University of Oklahoma Press, 2014.
Venable, Cecilia, Terry Palmer, Paul A. Montagna, and Gail Sutton. *Historical Review of the Nueces Estuary in the 20th Century: Final Report for Texas Water Development Board*. Corpus Christi: Harte Research Institute for Gulf of Mexico Studies, Texas A&M University–Corpus Christi, October 2011.
Wagert, Kam, and Pam Stranahan. *Aransas County in Postcards*. Fulton, TX: Friends of the History Center for Aransas County, 2014.
Wallace, Edward S. "General William Jenkins Worth and Texas." *Southwestern Historical Quarterly* 54 (October 1950): 159–68.
Watson, Richard. *Geologic Framework of the Eolian Sand Plain and the Central Flats of Laguna Madre and Circulation between Northern and Southern Laguna Madre*. Texas Coastal Geology. July 10, 2008. http://texascoastgeology.com/papers/giww_report.pdf.
Watson, Richard L., and E. William Behrens. "Nearshore Surface Currents, Southeastern Texas Gulf Coast." *Contributions in Marine Science* 15 (1970). https://texascoastgeology.com/papers/currents.pdf.
Weddle, Robert. *Changing Tides: Twilight and Dawn in the Spanish Sea*. College Station: Texas A&M University Press, 1995.
———. *The French Thorn: Rival Explorers in the Spanish Sea*. College Station: Texas A&M University Press, 1991.
———. "La Salle's Survivors." *Southwestern Historical Quarterly* 75 (1972): 425–26.
———. *The Wreck of the Belle, the Ruin of La Salle*. College Station: Texas A&M University Press, 2001.
Weems, John Edward. *A Weekend in September*. College Station: Texas A&M University Press, 1957.
Wegemman, Carroll H. *A Guide to the Geology of Rocky Mountain National Park (Colorado)*. National Park Service, 1995. http://www.nps.gov/parkhistory/online_books/romo5/wegemann/sec6.htm.
Weil, Harvey. "The History of the Port of Corpus Christi: 1926–2001." Port of Corpus Christi. https://portofcc.com/about/port/about-us/.
Welker, Glenn. "Karankawa Literature: The Karankawas." Indigenous Peoples Literature. http://www.indigenouspeople.net/karankaw.htm.
Werner, George C. "Texas Mexican Railway." *Handbook of Texas Online*. http://www.tshaonline.org/handbook/online/articles/eqt21.
"What Causes Hurricanes?" WeatherQuestions.com. www.weatherquestions.com.

White, Thomas J. *United States Early Radio History*. Blog. https://earlyradiohistory.us/sec005.htm.

White, William A., Thomas A. Tremblay, Rachel L. Waldinger, and Thomas R. Calnan. *Status and Trends of Wetland and Aquatic Habitats of Texas Barrier Islands, Coastal Bend*. Austin: Texas General Land Office, 2006. https://cbbep.org/publicaations/virtuallibrary/BarrierWetlandsCoastalBend.pdf.

Whitten, Horace Logan. "Marine Biology of the Government Jetties in the Gulf of Mexico Bordering the Texas Coast." Master's thesis, University of Texas, 1940. https://repositories.lib.utexas.edu/handle/2152/22109.

"WhoopingCrane." Encyclopedia.com. https://www.encyclopedia.com/people/history/historians-miscellaneous-biographies/whooping-crane.

Williams, C. Herndon. "The Star of St. Mary's of Aransas Never Went Out." *Baysider*, August 13, 2016.

———. *Texas Gulf Coast Stories*. Charleston, SC: History Press, 2010.

Wolff, Linda. *Indianola and Matagorda Island: 1837–1887*. Austin: Eakin Press, 1999.

Wolffarth, Louise. "The Transportation Problems of Corpus Christi, Texas." Master's thesis, Texas Technological College, 1939.

Wood, Alpha Kennedy. *Texas Coastal Bend: People and Places*. San Antonio: Naylor, 1971.

Worcester, Donald E. "Longhorn Cattle." *Handbook of Texas Online*. https://www.tshaonline.org/handbook/entries/longhorn-cattle.

INDEX

Adair, 72
Adams, L. M., 153, 154, 155, 156–57
Alabama, 30, 37
Alice Taylor, 68
Almonte, Juan Nepomuceno, 17
Anderson, Will, 137
Anna Hanson, 67
ants, 54
Aransas Bay: channel building, 85; colonist access difficulties, 23–24; hide-and-tallow factories, 59, 61–62; oyster beds, 77; shrimping activity, 165; Taylor's troops, 36; wharf construction impact, 41
Aransas Harbor Terminal Railway, 118–19
Aransas Pass Channel and Dock Company, 113, 118–19, 154
Aransas Pass Harbor Company, 96–97, 99, 100–101
Aransas Pass Terminal Railway, 137, 138, 147, 154
Aransas Pass (the town): commercial expansion, 125–26; deepwater port competition, 114–15, 156, 157, 162; hurricanes, 135–39, 144, 146–47, 148; naming of, 95; norther damage, 133; oil industry, 162–63; railroad benefits, 125–26
Aransas Pass (the waterway): blockade runners, 44; Civil War conflicts, 41, 42, 50–52; fishing activity, 122–25; jetty work map, *88*; military surveys, 39; passage barriers, 38; sediment movement patterns, 89–91; turtle harvests, 80; waterfowl hunting, 121–22
Aransas Pass (the waterway), alteration activity: cost statistics, 116, 208*nn*16–17, n21, 218*nn*11–14, 219*n*17; Darragh Brothers' project, 103–104; deepwater port competition, 112–18, 154–55; developer interests, 89, 93, 94–95, 99–100, 114–15; Ernst's project, 91–93, 107; Goodyear's project, 100–101; Kleberg's project, 104–105; Mansfield's project, 91, *92*, 102; maps, *88*, *90*, *92*, *98*, *104*, *109*; Picton's projects, 105–108; Ripley and Haupt's projects, 95–98, 101–103, *104*, 171; Ropes' project, 94; shifting sediments problem, 89–91, 101
Aransas River, 22
Aransas Road Company, 85
Aransas (steamer), 70, 199*n*28
Armada de Barlovento, 184*n*17
Armstrong, Clyde, 171
Army, U. S. *See* Civil War; Taylor, Zachary

Army Corps of Engineers, 83–84, 87, 91, 101–103, 112, 141, 151
Arthur, USS, 43–44
astrolabes, 13, 183*n*7. *See also* navigation aids
Austin, Stephen F., 17
Avarare people, 11–12, 26

Baffin Bay, 9
Banks, Nathaniel, 48
bar pilots, 64, 65–70, *71*, 169–72, 198*n*16, 221*n*1
barrier islands, overviews, *1–2*, 3–6, 175–78. *See also specific topics*, e.g., Aransas Pass *entries*; hurricanes; Padre Island; Rockport
Barstow, SS, 171–72
bascule bridge, 172–73
Baychester, 131
Beach, Lansing, 116
Bee, Carlos, 151
Bee, Hamilton, 49
The Belknap, *86*
Belle Italia, 45–47
Bérenger, Jean, 22–23
birds, 80–81, 111–12, 121–22, 126, *177*, 178
blockade attempts, Civil War, 41–42, 44
Boca del Rio, 58
Bolivar Peninsula, 72
Brazos River, 137
Brazos River and Dock Company, 85
Brazos Santiago, 44, 50, 58, 62, 85, 133
Breaker, 46
British Code List, 68–69, 199*n*26
Bromley, Carl, 171
Brownsville, 48, 49, 137
Buffalo Bayou, 85, 86, 156
Buffalo Bayou Ship Channel Company, 85

buoys, waterway, 68. *See also* navigation aids
Burges, Richard, 104
Burton, Isaac, 24

Cabeza de Vaca, Álvar Núñez, *8*, 11–12
Caller Times, 136
Camargo supply depot, 58
Capoque people, 11–12
Castillo, Alonso del, 12
cast nets, 78, 163
catarrhal fever, 33, 189*n*19
cattle industry, 41, 58–62, 66, 79, 87, 105, 167, 195*nn*7–8, 220n11
Cedar Bayou, 53, 71
channel building. *See* waterway alterations
Charles V, 10
Charruco people, 11–12
Civil War: coastal blockade plan, 41–44; Confederacy defenses, 45, 46–47, 50–52; Confederacy deserters, 44–45; fortifications map, *43*; pre-war Confederacy support, 39–40; Union attacks, 44, 45–46, 51–52, 192*n*15; Union invasion plan, 48–50, 51–54; Union supply ships, 58
Clendening, Frank, 138–39
Cline, Isaac, 134, 137
Clubb, Captain, 67
Code List of Signals, 68–69, 199*n*26
Coleman-Mathis-Fulton Company, 61
Colman, W. L., 136
Colonel Keith, 170
Comanche people, 26
compasses, 130, 183*n*7. *See also* navigation aids
Comstock, 117–18, 154
concrete ships, 129–30, 141

Confederacy, pre-war support for, 39–40
conscription, Confederacy, 44–45, 48
Copano Bay, *19*, 22, 23–24, 26, 41, 165
Copano people, 22–23
Copano (the town), 36
Cope, W. D., 147–48
Corpus Christi: breakwater construction, 151–52; Civil War activity, 46–48; commercial expansion, 126–27; disaster aid for Galveston, 134–35; hurricane attitudes, 136–37, 139–41; hurricane damages, 74, 138, 139–41, 144–46; port operations, 161–62, 169–74; seawall construction, 137, 151; Taylor's troops, 30–31
Corpus Christi, deepwater port competition: channel argument, 115; construction activity, 158–60; cost statistics, 116, 155–56, 208*nn*16–17, 218*nn*13–14, 219*n*17; designation decision, 157–58; design plan, 155, 156–57; lobbying activity, 112, 114, 127–28, 151–53; operations success, 161–62, 169–74
Corpus Christi, San Diego and Rio Grande Railway, 86–87
Corpus Christi Bay: Civil War fighting, 45–48, 192*n*15; Confederacy support, 39–41; European explorers, 24–25; hide-and-tallow factories, 59, 62; hurricanes, 145; lighthouse installation, 39; maps, *19*; protective barriers, 25–26; shoreline characteristics, 25; shrimping activity, 165; smuggler settlement, 26, 187*n*12; Taylor's troop activity, 24, 26, *27–29*, 29–34; waterway alteration, 85, 89, 94; wharf construction impact, 41

Corpus Christi Caller, 68, 87, 138
Corpus Christi Channel, 153, 155, 159
Corpus Christi Deep Water Committee, 152
Corpus Christi Pass, 38, 45
Cortés, Hernán, 10
Corypheus, 45–46
cotton shipments, 72, 118, 119, 156, 171
Craven, Lieutenant, 39
Creole, 37
cross-staffs, 10, 183*n*7
Cuba, 63
currents, 5–6, 13, 15, 89–90, 96, 97, 175–76

Dana, Napoleon, 33, 48–53, 132–33
Darlington, *129*, 130
Darragh Brothers, 103–104
depth-sounding tools, *14*, 68. *See also* navigation aids
disease, Taylor's encampment, 32, 33, 188*n*15
Dorantes, Andrés, 12
draft of ship, defined, 198*n*16
dredging activity: Corpus Christi port construction, 158–60; ecosystem damage, 41, 78, 97, 113, 175, 178; for Intercoastal Canal, 161. *See also* Aransas Pass (the waterway), alteration activity
Dred Scott decision, 39
drift currents, 5–6, 13, 15, 96
Dryden, SS, 171
dugout canoes, 7, 182*n*3
Dunn, Patrick, *55*, 60–61, 127, 141
Durham, 130

ecosystem damage: changes summarized, 175–79; Civil War attacks,

45, 53; dredging activity, 41, 78, 97, 113, 175, 178; La Salle's expedition, 12, 15; oyster beds, 77–78; railroad construction, 97; shrimping practices, 164–65; turtle harvests, 80; waterway dynamiting, 102; wharf construction, 41, 97
El Copano, 40
Elliot, Captain, 154
El Mar Rancho, 63–64
Emergency Fleet Corporation, 131
Encinal Peninsula, 60
Ernst, O. H., 90–92, 107
Escandón, José de, 25
Escondido River, 13, 24
Estevanico, 12
estuaries, fauna variety, 6

Farley, Barney, 168
Farley, Fred, 125
Farragut, D. G., 42
fauna: changes summarized, 176, 178–79; harvest practices, 76–82; hurricane impacts, 133, 143, 148; insect problems, 33, 54, 148; island variety, 6, 22; Taylor's troop encampment, 32, 33. *See also* fishing activity; hunting activity; shrimping industry
Fig people, 11–12, 26
fishing activity: commercial practices, 81–82, 126, 127; ecosystem damage summarized, 176, 178; native peoples, 7–9, 17, 22; recreational, 123–25, 127, 165–68, 210*n*5; Taylor's troops, 32
flag-and-light indicators, 68–70, 71. *See also* navigation aids
flies, 33
Follet's Island, 184*n*12

Fort Donelson, 42
Fort Esperanza, 48, 53
Fort Henry, 42
Fort Semmes, 50–53
Fort Worth Gazette, 89
Franciscan missions, 16–17
French arrival, 8, 12–15

Galveston Bay: LaSalle's expedition, 12; as Mexican port, 22; native lifeways, 9–10; shrimping activity, 78; waterway alteration, 83–84, 85
Galveston Bulletin, 77
Galveston Island, 22, 72
Galveston News, 102
Galveston (the town): hurricanes, 133–34, 137–38, 212*n*7; port operations, 71–72, 86, 87; seawall construction, 137, 214*n*15; waterway alteration, 83–84, 87, 93
Garner, John Vance, 104, 112, 114, 167
gathering activity, native peoples, 7, 9
geological processes, barrier islands, 3–6
Gillette, Cassius, 90
Goodyear, H. C., 100–101
Grant, Ulysses S., 30, 31, 33, 38, 41–42
grazing activity, 41, 60. *See also* cattle industry
Green, Ned, 125
Gresham, Walter, 104
grouper fishing, 123, 209*n*4
Guadalupe, 117
Gulf, Western Texas, and Pacific Railroad, 85–86
gunboats, Union, 42–44

Hall, William, 61
Hall's Bayou, 153, 155, 158

Index

Han people, 11–12
Harbor Island: developer interests, 101, 114; hurricanes, 141; petroleum industry, 162–63; railway terminus, 89, 93, 113; recreational visitors, 165–66; waterway alteration planning, 92, 95, 112, 113
Harbor Island Basin: construction of, 116–20, 212*n*2; hurricanes, 138–39, 141, 143–44, 154, 157; map, *109*; opposition to, 127, 128; shipping expansion, 119–20, 127; World War II impact, 128–29
Harmon, W. B., 141
Harrison, Russell, 94–95
harvest practices, coastal fauna, 76–82
Haupt, Lewis M., 95–98, 101–103, 171
hawsepipers, 169–72
Heldenfels Shipyards, 131, 141, 148, 163
Henry, William, 32, 33, 34, 37
hide-and-tallow factories, 55, 59, 61–62, 66, 79
high-profile barrier island, diagram, *5*
Hitchcock, Colonel, 38
Hobby, Alfred, 45, 46, 47
Hobby, William T., 131
horse marines, 24
Houston, 85, 156
Houston and Texas Central Railway, 85–86
Howell, George, 116
human occupation: Euro-American arrivals, 23–24; European arrivals, 10–17; native peoples, 7–10
Humble Company, 162
hunting activity: birds, 80–81, 121–22, 126; native peoples, 7, 9, 16–17; Taylor's troops, 32
hurricanes: attitudes about, 135–37, 139–41; and Corpus Christi's deepwater port, 156–57; damages from early 1900s events, 134–35, 137–48, 212*n*7; defined, 181*n*3; destructive capabilities, 133; map, *132*; Mercer's writing about, 74–75; sediment movement, 5

Indianola, 70, *74*, 75, 133, 134, 156
Indoe, James, 116
industries, map of, *161*
Ingleside, 162
Ingram, J. S., 124
insect problems, 33, 54, 148
Intercoastal Canal, 161. *See also* waterway alterations
Isla Blanca, 24
Isla de Malhado, 11, 72, 184*n*12

Jervey, Henry, 116
jetty building. *See* waterway alterations
John Jacobson, 159–60
Johnson, Charles, 128

Kahl, Nick, 141
Kansas-Nebraska Act, 39
Karankawa people, *8*, *9*, 12, 15–17, 26, 183*n*5, 185*n*23
Kenedy, Mifflin, 58–59, 86–87
King, Henrietta Chamberlain, 89
King, Richard, 58–60, 86–87, 195*nn*7–8, *nn*12–13
King Ranch, 55, 105
Kinney, Henry L., 26, 30–31
Kittredge, John W., 44–48, 85, 192*n*15
Kleberg, Robert, 105, 112, 148, 151, 152, 153
Kleberg, Rudolph, 105

La Baca, 40
Laguna Madre, 4, 36, 49, 61, 85, 124, 145, 163, 165
La Salle expeditions, *8*, 12–15, 133
Lavaca, 70–71
lead lines, 10, *14*, 36, 68, 183*n*7
León, Ponce de, 16
lifesaving station, Mustang Island, 73
lightering, defined, 188*n*5
lighthouses, 39
littoral currents, 89–90, 96, 97
livestock, ecological impact, 41
longhorn cattle, characteristics, 57–58. *See also* cattle industry
Lott, Uriah, 86
Lovenskiold, Colonel, 85
low-profile barrier island, diagram, *5*
Lubbock, Governor, 44
Lucas, Cyrus B., 220*n*11
Lydia Ann Channel, 117, 155
Lydia Ann Point, 67

MacDonald Engineering Company, 129–30, 141
Magruder, John, 51
Malhado Island, 11, 72, 184*n*12
Mallory Shipping, 86
Mann, Billy, 46
Mansfield, Samuel M. (and jetty), 91, *92*, 102, 117
Marconi, Guglielmo, 130
Marcy, William L., 29, 36
Mariame people, 11–12
Marion Packing Company, *62*
martial law, 44
Mary, 72–73
Matagorda Bay: European arrivals, 11, 12–13, 15; hurricane damage, 70, 141; reconnaissance team, 36; waterway alteration, 84–85
Matagorda Island, 11, 12–13, 15, 36
Maury, Matthew F., 134
McClellan, 51
McClelland, George, 38–39
Meade, George, 33
Mercer, Agnes, 62
Mercer, Edward T. "Tom," 103, 120, 135, 141, 170–71
Mercer, John, 66, 72–73
Mercer, Ned, 66, 67, 72
Mercer, Peter, 63
Mercer, Robert Ainsworth, 41, 62–63, 65–66, 69, 73–74
Mercer, William Henry, 63
Mercer (not named), notes by, 67, 68, 72, 74–75
Mercer settlement, 41, 44, 94
Mexico, Velasco Treaties, 27–28, 187*n*1. *See also* Santa Anna; Taylor, Zachary
Mier supply depot, 58
Miller, Roy, 114, 127–28, 151–52, 153, 158
Miruelo, Diego, 10–11
Mission River, 22
Mississippi River, 13
Moctezuma, 22
Monongahela, 51
Monroe, Captain, 39, *40*
Montgomery, Marshall, 54
Morgan, Charles, 62, 66, 84, 85–86
Morris and Cummings Cut, 97, 105, 107
Morse Code, *172*
mudshell, 77–78
Mustang Island: birds, 111, 121–22; Civil War activity, 44, 45, 50, 51–52, 53–54, 63; as Corpus Christi

protection, 26, 136–37; European arrivals, 11, 13; hurricanes, 135, 138, 143; Mercer's appreciation for, 64–65; Mercer's bar piloting, 63–66; Mercer settlement, 41, 44, 94; packing house, 59; waterway alteration, 89, 91, 93–94, 95, 98, 105; waterway navigation aids, 39, 68–69. *See also* Aransas Pass (the waterway), alteration activity

Narváez, Pánfilo de, 10–11, 132
native people: European contacts, 11, 15–16; pre-European lifeways, 7–10, 22–23
navigation aids, 39, 64–70, 130, 172, 183*n*7. *See also* bar pilots
navigation districts, creations, 152–53, 155, 158, 163
Navy, U.S., 39, 42–48
nets, harvesting, 78, 80, 82
New City, plans for, 89, 93, 94–95. *See also* Aransas Pass (the town)
Newcomer, H. C., 155
New York Times, 134, 142
Nicaragua, 137
nocturnals, 10, *14*, 183*n*7. *See also* navigation aids
northers: defined, 181*n*3; Haupt's breakwater design, 96; La Salle expedition, 15; Mercer's observations, 73; Morgan steamship, 72; Savage's embankment, 93; and sediment movement, 4, 90, 96; Taylor's troops, 33, 36; Union troops, 53. *See also* hurricanes
Nueces Bay, 24, 25, 127, 140–41, 145–46, 151, 155

Nueces River, 24, 28, 36, 41, 127
Nuestra Señora del Refugio, 16–17

oil industry, 119, 129, 141, 154, 171–72
Ol' South Tanks, 60
Oppikoper, F., 103
Orobio y Basterra, Joaquin de, 24–25
Ortiz Parilla, Diego, 24–25
Oso Creek, 59
otter trawls, 163–64
oyster harvesting, native peoples, 7, 9, 23
oyster reefs, as ship barriers, 23, 25
oyster shells, road construction, 77–78

Packery Channel, 148, 153, 165
packing houses, 55, 59, 61–62, 66, 79
Padre Island: Civil War activity, 45, 48, 51, 58, 194*n*31; as Corpus Christi protection, 26; hurricane damages, 137, 141, 142–43, 148; King's holdings, 59, 60, 195*nn*12–13; naming of, 24; post-hurricane recovery, 153–54; waterway alteration, 85, 88–89
Pass Cavallo, 71
petroleum shipments, 119, 129, 141, 154, 171–72
Philadelphia Times, 75
Picton, D. M., 105–108
Pilot Boy, 139
pilots, ship, 10–11, 64, 65–70, 169–74, 198*n*16
Pineda, Alonso Ávarez de, 24, *25*, 186*n*9
Pleasant, Alclair Mays, 146
Pope, W. E., 158
Port Aransas: deepwater port competition, 116, 135; growth of, 125; hurricanes, 139, *140*, 141, 143–44, 146,

151–52; industry wages, 163, 220*n*3; petroleum shipments, 162–63; post-hurricane recovery, 154; sport fishing, 122–24; tourism industry, 166–68

Port Aransas Post, 116

Port Arthur, 138

Porter, John B., 33

Port Isabel, 48, 49, 137

Powers Southern Dredging Company, 113

prehistory, 3–6

Providence, 154

Quaker guns, 51, 194*n*30

radio technology, 130, 171

railroads: deepwater port connections, 118–19, 155, 156; ecosystem damage, 97; and Ernst's waterway alteration, 93; hurricane damages, 139, 141, 145, 147, 154; recreational travel, 165–66. *See also* San Antonio and Aransas Pass Railroad

Ramón, Domingo, 16

rattlesnakes, 33

reconnaissance teams, Taylor's, 35–36, 39

recreational fishing, 123–25, 127, 165–68, 210*n*5

Redfish Bay, 126

Reeder, George, 136–37

Reindeer, 45–47

Reynolds, Alfred, 48

Reynosa supply depot, 58

Richardson, Sid, 167, 168, 220*n*11

Riche, Charles, 116

Rio de las Palmas, 10–11, 184*n*10

Rio Grande, 28–29, 58

Ripley, H. C., 94, 95–98, 101–103

Rivers and Harbors Committee, 105, 157–58

road construction, oyster shells, 77–78

Rockport: decline in federal dredging interest, 162; disaster aid for Galveston, 135; growth of, 128–29; hurricanes, 139, 144, 146–48; shipbuilding industry, 130–31, 141, 163; shipping activity, 61–62, 66; shrimping industry, 128, 163–65; waterway route to, *64*

Rockport, deepwater port competition: arguments for, 89, 112, 115; cost statistics, 116, 208*nn*16–17, 218*nn*12; design plans, 155; financing plans, 155, 156; hurricane argument, 157

Roosevelt, Elliott, 167

Roosevelt, Franklin D., 167–68

Ropes, Elihu, 89–90, 94

Sabine Pass, 48, 133–34

Sachem, 45, 46–47

Sam Houston, 141

San Antonio and Aransas Pass Railroad: in Aransas Pass town, 125–26, 155; in deepwater port plans, 155; formation, 86–87; hurricane damages, 139, 141, 145, 148; locomotive, *86*; New City's plans, 94–95; Padre Island developer expectations, 88–89; Rockport's plans, 112; route map, *84*. *See also* railroads

San Antonio Bay, 26

San Antonio Chamber of Commerce, 152

San Antonio Express, 143

San Antonio Progress, 102

San Carlos de los Malaquittas, 24

sand bars, 23–24
San Miguel Arcangel, 25
Santa Anna, Antonio López de, 22, 24, 27–28
Savage, J. E., 93
schooners, advantages, 21–22
Sciota, 48
scorpions, 33
seafood processing plants, *76*
sea-grass beds: dredging/construction damage, 41, 97, 113, 175, 178; fish populations, 121, 122, 178; hurricane damage, 133; sea turtle populations, 79, 80; soldier damage, 12
sea turtles, *76*, 78–80, 176, 200*n*7
secession plans, Texas, 39
sediment movements, overviews, 3–6, 89–91, 95–96, 176. *See also specific topics*, e.g., Aransas Pass (the waterway) *entries*; currents; winds
seine nets, 78, 82, 163
Seminole Wars, 58
shallops, 12, 184*n*15
Shell Bank Island, 44, 45, 194*n*31
shellcrete, 78, 85
Shell Island, 38
ship-building industry, 129–31, 141, 163
Ship Channel, Houston's, 156
shipping companies, 58–59, 62, 66–67
ship technology, changes, 21–22, 119–20, 129–30
shrimping industry, 78, 163–65, 168, 176
Sibley, Henry Hopkins, 41, 190*n*15
Sigmund, John, 162
signal points, waterway, 39, 65, 67–68, 172. *See also* navigation aids
Signal Service Corps, 73
single reaction breakwater, Ripley and Haupt's, 95–98, 101–103, 171

Slayden, James, 114–15
smuggler settlement, Corpus Christi Bay, 26, 187*n*12
smuggling activity, 22, 26, 44
South and West Texas Deep Water Harbor Association, 152
Spanish arrivals, *8*, 10–11, 13, 16
St. Joseph Island: birds, 111–12; cattle industry, 167; Civil War conflicts, 53; Copano Bay access, 23–24; Copano people, 22, 23; as Corpus Christi protection, 26; FDR's visit, 168; hurricanes, 74, 138, 144, 220*n*11; sales of, 220*n*11; signal points, 39; Taylor's troops, 30, 36, 37, 38; waterway alteration, 91, 92, 95, 116, 212*n*2
St. Mary's, 41, 62
steamboats, 37, 38, 58–59
Stillage, H. S., 39
Stow and Company, 77
sundials, *14*
Supreme Court ruling, bar pilots, 64
Surprise, 72

tariff regulations, Mexico's, 22
tarpon fishing, 123–25, 127, 165–68, 210*n*5
Tarpon (the town), 115–16
Taylor, H., 156
Taylor, Zachary: coastal reconnaissance teams, 35–36, 39; encampment at Corpus Christi, *28–29*, 30–34, 188*n*15; supply depots, 58; travel to Corpus Christi Bay, 26–27, 29–30, 37–38
Tedford, W. E., 146
Tennessee, 49–50
Terán de los Ríos, Diego, 16
Texas Mexican Railway, 88

Texas Sun, 99–100
Throckmorton, James Webb, 66
tourism industry, 122–24, 128, 166–68
trail drives, 58, 59, 61. *See also* cattle industry
Traylor, Harry, 128
tugboats, 119–20, 170–71
Turtle Bay Resolutions, 22
Turtle Cove, 105, 112–13, 117, 127, 155
turtle processing plants, *76*
turtles, sea, *76*, 78–80, 176, 200*n*7
Twigg, David Emanuel, 30, 32

Undine, 37
Upton, Bern, 32

vegetation: ecosystem damages, 41, 176; island variety, 6, 22, *23*, 25, 60. *See also* sea-grass beds
Velasco Treaties, 27–28, 187*n*1
von Blücher, Felix, 46
von Blücher, Maria, 192*n*15

waterfowl hunting, 80–81, 121–22, 126
water supplies: Padre Island, 60; Taylor's troop encampment, 33
waterway alterations: Galveston town, 83–84, 87, 93; Harbor Island Basin construction, 116–20, 212*n*2; Matagorda Bay, 84–85; Padre Island, 85, 88–89. *See also* Aransas Pass (the waterway), alteration activity; Corpus Christi, deepwater port competition
Welles, Gideon, 47, 48
Wheeler, T. B. "Tom," 94–95, 103–104
whiskey story, shell duds, 192*n*15
Whisper, 67
white supremacy, 39
Whiting, Daniel, 32, 33
wildlife. *See* fauna
Willett, John, 88–89, 100
winds, 4–5, 33, 73, 90, 96, 132–33. *See also* hurricanes; northers
Wood, Leonard, 123
Wood, William, 61, 220*n*11
World War I, 128–29, 131
wreck master responsibilities, 73. *See also* bar pilots

Yriarte, Pedro de, 16

Zuniga, 131

OTHER BOOKS IN THE GULF COAST BOOKS SERIES

Lighthouses of Texas
LINDSAY T. BAKER AND F. R. HOLLAND

Laguna Madre of Texas and Tamaulipas
JOHN W. TUNNELL AND FRANK W. JUDD

Fishing Yesterday's Gulf Coast
BARNEY FARLEY AND LARRY MCEACHRON

Designing the Bayous: The Control of Water in the Atchafalaya Basin, 1800–1995
MARTIN REUSS

Life on Matagorda Island
WAYNE H. MCALISTER AND MARTHA K. MCALISTER

Book of Texas Bays
JIM BLACKBURN AND JIM OLIVE

Plants of the Texas Coastal Bend
ROY L. LEHMAN AND TAMMY WHITE

Galveston Bay
SALLY E. ANTROBUS

Crossing the Rio Grande: An Immigrant's Life in the 1880s
LUIS G. GÓMEZ AND GUADALUPE VALDEZ

Birdlife of Houston, Galveston, and the Upper Texas Coast
TED EUBANKS AND ROBERT A. BEHRSTOCK

Formation and Future of the Upper Texas Coast: A Geologist Answers Questions about Sand, Storms, and Living by the Sea
JOHN B. ANDERSON

Finding Birds on the Great Texas Coastal Birding Trail: Houston, Galveston, and the Upper Texas Coast
TED EUBANKS AND ROBERT A. BEHRSTOCK

Texas Coral Reefs
JESSE CANCELMO AND SYLVIA EARLE

Fishes of the Texas Laguna Madre: A Guide for Anglers and Naturalists
DAVID A. MCKEE AND HENRY COMPTON

Louisiana Coast: Guide to an American Wetland
GAY M. GOMEZ

Storm over the Bay: The People of Corpus Christi and Their Port
MARY JO O'REAR

After Ike: Aerial Views from the No-Fly Zone
BRYAN CARLILE

Kayaking the Texas Coast
JOHN WHORFF

Glory of the Silver King: The Golden Age of Tarpon Fishing
HART STILWELL

River Music: An Atchafalaya Story
ANN MCCUTCHAN

Del Pueblo: A History of Houston's Hispanic Community
THOMAS H. KRENECK

Letters to Alice: Birth of the Kleberg-King Ranch Dynasty
JANE CLEMENTS MONDAY AND FRANCIS BRANNEN VICK

Hundred Years of Texas Waterfowl Hunting: The Decoys, Guides, Clubs, and Places, 1870s to 1970s
R. K. SAWYER

Texas Market Hunting: Stories of Waterfowl, Game Laws, and Outlaws
R. K. SAWYER

Fire in the Sea: Bioluminescence and Henry Compton's Art of the Deep
DAVID A. MCKEE

Pioneering Archaeology in the Texas Coastal Bend: The Pape-Tunnell Collection
JOHN W. TUNNELL AND JACE TUNNELL

Vertical Reefs: Life on Oil and Gas Platforms in the Gulf of Mexico
MARY KATHERINE WICKSTEN

Glorious Gulf of Mexico: Life Below the Blue
JESSE CANCELMO

Alligators of Texas
LOUISE HAYES AND PHILIPPE HENRY

Bulwark Against the Bay: The People of Corpus Christi and Their Seawall
MARY JO O'REAR

Texan Plan for the Texas Coast
JAMES B. BLACKBURN

Dr. Arthur Spohn: Surgeon, Inventor, and Texas Medical Pioneer
JANE CLEMENTS MONDAY

Fishes of the Rainbow : Henry Compton's Art of the Reefs
DAVID A. MCKEE

Protecting Historic Coastal Cities: Case Studies in Resilience
EDITED BY MATTHEW PELZ